Joseph Parrish

Alcoholic inebriety

from a medical standpoint

Joseph Parrish

Alcoholic inebriety
 from a medical standpoint

ISBN/EAN: 9783744737449

Printed in Europe, USA, Canada, Australia, Japan

Cover: Foto ©berggeist007 / pixelio.de

More available books at **www.hansebooks.com**

ALCOHOLIC INEBRIETY:

FROM

A MEDICAL STANDPOINT.

WITH

CASES FROM CLINICAL RECORDS.

BY

JOSEPH PARRISH, M.D.

PHILADELPHIA:

P. BLAKISTON, SON & CO.,

1012 WALNUT STREET.

1883.

TO THE MEMORY OF

DONALD DALRYMPLE, M.P., M.D., J.P.,
D.L. for the County of Norfolk,
F.R.G.S.,

AS A TOKEN OF THE HIGH PERSONAL ESTEEM IN WHICH HE

WAS HELD IN THE

UNITED STATES OF AMERICA,

AND IN REMEMBRANCE ALSO OF HIS DISTINGUISHED

SERVICES AS CHAIRMAN OF THE

SELECT COMMITTEE OF THE HOUSE OF COMMONS

ON

HABITUAL DRUNKARDS,

THIS VOLUME IS RESPECTFULLY DEDICATED

BY THE AUTHOR.

his surroundings, to attempt to penetrate tissues, and search after forces with which they are not familiar.

It has been my purpose, therefore, in the following pages, to state principles and facts, and enforce them, so far as I could, from the material in my possession, by cases—all of which have either been under my own care, or within my knowledge.

That physicians may be induced to take a deeper interest in this subject, if examined from their own standpoint, is a hope which has stimulated my endeavor to present the aspect of disease, especially as it is fortified by the testimony of some of the most eminent men of the profession.

Those physicians who are specially interested in nervous diseases, and whose attention to morbid psychology has enabled them to contribute freely to advance the science of psychiatry, may find abundant aids to their researches by studying from their special standpoint the disease we are considering.

Following these thoughts, I have allowed them to shape the plan of the book, which, though it might be improved as to arrangement, is nevertheless the vehicle for presenting what I regard as highly important truths.

To the medical profession I would say that the sections on Traumatic Inebriety, on the Trance State, and on the Relation between Insanity and Inebriety, discuss topics of increasing importance and interest, especially in a medico-legal sense, and should be freely investigated.

THE AUTHOR.

Burlington, New Jersey, April, 1883.

CONTENTS.

CASES TO BE FOUND IN THE TEXT

ILLUSTRATIVE OF DIFFERENT FORMS OF INEBRIETY.

ALCOHOLIC INEBRIETY

FROM A MEDICAL STANDPOINT.

WHO ARE INEBRIATES?

There are some persons who will never be drunkards, and others who will be so, in spite of all that can be done for them. Some are drunkards by choice, and some by necessity.—*Macnish—Anatomy of Drunkenness.*

In the present state of knowledge, we have no need to retrace steps that have already been taken, to demonstrate the universal prevalence of a desire for artificial stimulation among all peoples, in every nation on the globe. It is sufficient to recognize the fact, that while there is a natural and necessary adaptation of our standard foods to the normal demands of the human economy in a state of health, there is an equally apparent adaptation of the distinct and separate chemical qualities of such foods, to the various abnormal conditions which afflict humanity in the form of disease, or which disclose themselves in that wide range of physical disturbances in which there is neither perfect health, nor localized disorder. This fact constitutes the basis of our materia medica, and is the starting point of the science of therapeutics. Within this limitless range of morbid feelings, and undefined sensations, we lose sight of the border line between health and disease, and so far as the selection of remedies is concerned, the distinction between foods and medicines is also obliterated. What are recognized by all, as foods and medicines are employed interchangeably in the treatment of the multitude of cases which are neither sick nor well. It is among such persons we must look for the consumers of alcoholic beverages, and such drugs as possess similar qualities. Whether such qualities are to be regarded as nutritive or medicinal, depends of course, upon the degree of departure in each case from the typical standard of perfect health. Such persons are driven by their necessities, to explore the region of artificial, or more properly, unnatural nutrients, when they begin to realize a craving or lust to which they have hitherto been strangers. As the vital chemistry within them, fails to extract from natural and wholesome food, suitable nourishment for their enervated or perverted nerve structure, they search after such substances as the chemistry of science may evolve from the fruits of the earth, that they may be composed, exhilarated or narcotized, in accordance with the concealed cravings within.

Evenly balanced people, who are by nature, calm and self-possessed, and in good bodily health, are not those who usually fall into excess. With bodies in a normal state, and fed by natural food; and minds well poised, deliberate and disciplined by culture, thus constituting a being self-controlled and vigorous, they have no need to venture into the realm of artificial nutrients, seeking for specifics to recuperate exhausted nerves, or to supple-

ment ordinary diet. We must look for the chief factors
of the craving for drinks among the ailing and half
sick, who suffer from disquietude of nerve, dyspepsia,
and the various hysteric and kindred phenomena, that
are now so readily recognized, even by unprofessional
observers.

Thus we come naturally and logically to apprehend
the remote causes of inebriety. That they are intrinsic
and belong to the individual, is without doubt, true.
This brings us also to the real issue that is involved in
the subject we are considering. It is not a legal question, the issue being revenue, or no revenue. It is not
simply a moral question, the issue being between the
use, and non-use of intoxicants. It is a question of
nerves—a neurosis—the issue being between soundness,
and unsoundness of structure or function; between a
complete and an incomplete manhood. It is disease,
and in the language of the "American Association for
the Cure of Inebriates," "a disease that is curable in the
same sense that other diseases are, its primary cause
being a constitutional susceptibility to the alcoholic impression, which may be inherited or acquired."

This disease, however, is not to be regarded as an
entity that approaches and invades the human organism
from without, but rather as a variation of natural function, having its source in the system itself. It may be
implanted somewhere in the complex structure which
constitutes the man, by hereditary taint. It may be, by
some obscure and undefined qualities coming together,
which, by their active and retroactive processes, originate

and evolve symptoms that indicate a departure from a healthy standard.

The existence of a predisposition to physical disease and to moral and mental qualities, to say nothing of resemblances of persons and manners, is a fact which is as familiar to the people, as any other fact in the natural history of the race. It constitutes a part of each family record, and belongs, as an entail, to the inheritance of every household. In its relation to the subject before us, no exception can be made. The law of heredity is inflexible, and its behests are without compromise.

There are, however, aspects of the subject which are recognized by the popular verdict as vice and crime, that deserve a share of attention, before proceeding to discuss the subject in its aspect as disease.

THE VICE ASPECT.

By inebriety as a vice, is meant that form of it which is not characterized in the beginning, by any noticeable physical longing, or deep-seated appetite or craving, such as is experienced by those whose nerve centres are disturbed, either by external injury, or by latent functional impulse. It is a form of inebriety resulting primarily from a mental or emotional prompting to indulge in intoxicants. There are multitudes of persons who do not consult either conscience or judgment, but who are ready to tread any path that others may mark out for them, if it leads in the direction of bodily or mental pleasure.

Such persons not unfrequently drift into evil practices, without seeming to be aware of the course they are pursuing. To them, the first "drunken fit" is an accident. Having once fallen, especially if the accessories of the debauch were at first pleasing, it is very easy to float down the same enticing current far enough to create and establish the very condition, that eventuates in the confirmed neurosis which constitutes the disease. At first it is a vicious propensity, indulgence in which may terminate in disease or crime, or both. It has its counterpart in numerous other departures from the laws of health, which terminate sometimes in disease, and sometimes in criminal conduct, and the analogy between this disease and others, both as relates to causation, progress and results, cannot be maintained without the presentation of this view of the subject.

A person may be so regardless of the laws of health as to indulge in a course of diet, which, though proper in moderation and at suitable intervals, may be injurious under opposite circumstances, finally creating disorder of the alimentary system, and resulting in chronic disease. The primary act of inordinate use may be a mere vice, which leads on to repeated vicious indulgences, the outcome of which is a fixed pathological state. As applied to intoxicants, as well as to ordinary food, the term vice may be legitimately employed in its application to the primary acts, as justly as the term disease may be applied to the developed condition, that is recognized as pathological.

It is somewhat difficult to determine the dividing line between vice and disease, and yet there is a certain pathognomonic sign, which, if it has not been accurately described by others, is yet, I think, capable of delineation as marking the advent of a disordered state; and that is,—a change of character. After repeated indulgences in the use of alcoholic liquors, so as to establish the habit, and a craving for its continuance is set up, the character becomes modified, and the phenomena of disease manifest themselves.*

Dr. Forbes Winslow, of London, the late distinguished author and teacher, says:—

"There is a normal drunkenness, as there are normal forms of any other vice. It is very difficult to deal with drunkenness as a vice, but when it passes the boundary, and ceases to be a vicious propensity, then the morbid craving for stimulants is clearly traceable to the mental condition. * * * The boundary line may be drawn between the vice of gluttony and the diseases which may follow, in which visceral inflammation and deterioration take place, and various forms of organic disease are developed."

The following case is offered in illustration :—

No. 460. A. B., æt. 30. A wayward, impulsive youth, with no regular employment. He lacks application and industry, and is fond of mirthful associates, and

* My friend, Dr. T. D. Crothers, refers to this fact in several cases recorded in his pamphlet on "The Trance State in Inebriety." A railroad conductor who drank moderately for two years, when "his disposition began to change." He became "more suspicious as he grew older," etc., "his mind exhibited evidence of failure in the irritability and changed disposition, etc."

ready for excitement of any description. A wide circle of acquaintances visit his father's house, and he is always ready to join them, at home or abroad, in any engagement that promises either mirth or sorrow. He was occasionally led into intoxication, which was generally, pleasurable to him. Recovering from it, and being reproved by his family, he would naturally swing over to the other extreme, and it was quite common for him to be seized by paroxysms of religious fervor, when he would sign the pledge, attend religious meetings, and abstain for months at a time. He would pursue such a life under the leading of his mother and sisters, and then suddenly break away again, being at the time excited by an election, a fire, or something that caused a sudden deviation from his steady line of conduct. The deflection once begun, originated a new course in the direction toward which it naturally led, till, enfeebled by its own persistency, it was readily exhausted, and another departure from that course, created still another line of divergence, which, followed to its end, completed the zig-zag outline of his character. I have known him to watch by the sick, night after night, administer stimulants punctually, without tasting them himself, and perform all the offices of a nurse, with the tenderness and delicacy of a woman. On the occasion of a death, he would wash the cadaver, and apply himself, of choice, to needless menial services, and after all was over, weep with the apparent grief of a brother, whether for the loss of a hospital companion, or for the loss of an opportunity to distinguish himself, I know not. In the absence of

such or similar occasions, when body and mind were
fully occupied, I have known him debauched to the
lowest degree, and afterward be as repentant and devout
as before. His life was made up of self-sought, if not
self-created tragic displays. His drunkenness was a
mere incident, which took its turn with other vices, as
occasion or opportunity might determine. Such men
frequently return to a life of sobriety and uprightness,
if, in the race between the worse and better qualities of
their nature, the latter may maintain the ascendancy
long enough to fix the habit, and give permanency to the
inclination for a steady life. If, however, the deflec-
tions from a normal line are toward the evil side of life
more frequently, and for longer periods, than toward the
virtuous and good, the evidences of disease become
apparent. Among these may be named, irritability of
temper, a spirit of exaction, a self-consciousness that is
offensively exaggerated, a suspicion that is morbid, a
degree of secrecy and falsehood, that was not previously
exhibited, together with boastful and pretentious language; all of which may be grouped with similar symptoms, as pathognomonic of disordered cerebral function.
The man is changed. He is no longer a wayward youth,
with character scarcely formed; but he is a victim of
disease, that is marked by positive symptoms, that fix
his character, and establish him as an inebriate.

Another case of like character has come under my
notice, though not under my care, but it is so distinctly
emotional in type, that I cannot forbear noticing it
briefly :—

For several months of the year he is an inmate of an inebriate's ward of a public institution. Committed for drunkenness only, guilty of no crime, and creating no disturbance, but simply docketed as "drunk." During other months of the year, when religious revivals and temperance lectures are in order, he is actively engaged in good works. He is reported to be an efficient temperance advocate, and is sometimes employed as a public speaker in behalf of the cause. When intoxicated, he is lively and joyous, and a religious fervor of expression and devout performances simulating worship characterize his debauchery. He is said to be free from other vices, and in the indulgence of the one, evinces no taste or tendency in the direction of profanity, lewdness, or any sort of crime. He will probably terminate his life in an asylum, though it may be that such a change may occur in his cerebral organism, as to give a wholesome direction to his tastes.

Dr. Winslow attributes the morbid craving to a new mental condition, and, of course, the reverse is as true, that a cessation of the morbid craving, is due to an altered mental condition. A significant fact in connection with waste of brain substance is, that it is repaired, according to Carpenter, "by a complete reproduction of normal tissue." Now, if it be true that the condition of the brain that determines human conduct in any given direction, must be modified in order to produce a different line of conduct in an opposite direction, it is easily understood how, by the substitution of new tissue for the old and wasted substance, a new direction may be given to the life and conduct of the individual.

The tendency in nature being toward the maintenance of the perfect type, we may look for an endowment of new normal tissue where all the conditions are favorable, and under such circumstances a cure, or what is popularly called reformation, takes place. Around this single fact are clustered the opposing theories and statements which characterize the history of this subject. If the public mind could lay hold of, and appreciate the doctrine that a sound physiological basis is essential to a perfectly sound and evenly balanced moral nature, there would not be the degree of divergence between the real truth, and what is commonly accepted as truth, in this matter. In the most common forms of alimentary disorder it is not difficult for people to observe the altered *morale* to which reference has already been made. The boundary line between the vice of gluttony, and the diseases which follow it, in which there is , visceral inflammation and deterioration, is readily distinguished by the very same class of symptoms that designate the disease of inebriety. As the vice passes into a state of visceral derangement, which is disease, the change of character begins to manifest itself. Nothing is more common, than the irritable temper of dyspeptics, the gloomy moodiness that accompanies liver derangement, or the odd fancies, and vagaries of hysteria, which so commonly represent sympathy with diseased organs. The same law is applicable to the subject in hand, and no one who has observed closely, can have failed to notice the differences of character which are exhibited in the career of an inebriate, as he progresses from the careless, motiveless beginning, to the stage of cerebral disorder.

If there is one prevailing symptom which is common alike to all, it is an ever present and magnified consciousness of self. Such persons require attention, and exact it of others. They are self-important, and demand a recognition of their importance by others. Their symptoms are exaggerated, and their sufferings intense, and unless this is appreciated by others as by themselves, they are provoked, and sometimes passionate. These symptoms do not appear, however, till the boundary line is passed.

Reviewing the vice aspect of inebriety, there seems to be some ground upon which to rest the theory that it is not always disease, though Dr. T. D. Crothers, of Hartford, Conn., who has had ample opportunities for observation and study, writes to me as follows:—

"Intemperance as a vice is not a clinical reality, and this term will be dropped when we know more about the subject. I think it is misleading, and could be applied with as much propriety to the first stages of paralysis, and to many cases of insanity, when the first symptoms are changes of character, morals, habits, etc. My view that inebriety has *never a stage of vice*, is confirmed by authorities abroad. Dr. Boddington, of England, is emphatic in calling it a disease from the time of the first narcotic or toxic action of alcohol."

If this view is accepted, and the boundary line is made the *toxic action* of alcohol, then the position is the same

as that occupied by those who recognize the vice as
ante-dating the use of alcohol. It is important, however,
to appreciate the fact, that there is a point where the
poison of alcohol produces its specific effect, and that the
craving for drink is coincident with this crisis in the
career of the individual. The desire now comes from
a physical cause; a toxic impression has been made upon
the organism in some part, and its effect is to reproduce
itself, and intensify its demands.

In connection with this branch of our subject, and in
seeming confirmation of the views of Dr. Crothers, I
may state that, in a recent review of my own clinical
records, I find an almost constant and uniform coincidence of disorders of various kinds, which I originally
marked as only *complications*—hernias, hemorrhoids,
strictures, venereal symptoms, gastric and hepatic troubles, rheumatism, neuralgias, cystitis, chronic ulcers, and
a variety of minor accidents and deformities—but at the
time I had no thought of entering them upon the record .
as having any causal relation to inebriety in either of
its forms. Not until my attention has been more
recently called to it, have I thought of correlating these
several conditions with the fact of intoxication, in any
other than an accidental relation. I am, however, more
and more disposed to study them together, as there is
probably a definite, though unobserved dependence upon
each other.

Dr. J. Milner Fothergill, of London, in a letter to me
recently, in relation to insanity, thus expresses himself.

"I am familiar with a case where wild delirium is

the accompaniment of acute indigestion." In a paper he has kindly sent me to read before the "National Association for the Prevention of Insanity," the same views are strongly presented.

Assuming that gout is due to an excess of lithic and uric acids in the blood, and that this vitiated and poisoned blood, penetrating and accumulating in the cerebral vessels, produces brain disturbance, and of course unusual conduct, which is not unfrequently pronounced insanity, why may not the insanity-tendency exhibit itself in the propensity to drink to excess, the drinker being unconscious at the time of any relation between his morbid diathesis, whatever that may be, and his habit of intoxication?

The same may be said of biliary products, which are proverbially connected with mental gloom and depression, and it would seem that any irritating or disturbing centre, located where it may be, would be competent to awaken either a psychical impulse to drink, to relieve morbid sensations, or to disclose a brain affection, which in, and of itself, predisposes to intoxication. What is true of gout and allied morbid conditions, in their relation to insanity, may be equally true as to inebriety, and if so, it would seem improper to diagnose simple acute drunkenness without reference to the history of the individual. It is but just to award to the inebriate whatever his history, and constitution may demand, in the way of extenuation.

Dr. Fothergill further says that "the brain can no more do its work properly and smoothly, when poisoned by uric acid, than the patient can have an elastic tread with a gouty foot." He goes so far in harmonizing the symptomatology of gout and insanity, as to place them side by side, in their consequences as to responsibility of conduct, and says, in referring to irregular conduct, "It is the perversity and cantankerousness of a gouty phase of mind—gout in the brain indeed—and the recording angel will probably make a note to that effect, for a plea of extenuating circumstances at the last Assize."

I conclude the consideration of the vice aspect by the following quotations, most of which are taken from the reports of a Select Committee on Habitual Drunkards, to the House of Commons of England, 1872.

Dr. Forbes Winslow, M. D., D. C. L., Oxon, a distinguished teacher and author, says:—

"There is an enormous mass of drunkenness in the lower classes which cannot be traced either to mental or brain disease, in the right acceptation of these words."

*   *   *   *   *   *   *   *   *

"I think there are habitual drunkards, as there are habitual prostitutes, and persons who habitually indulge in any other form of *vice*. It is their natural and normal state."

*   *   *   *   *   *   *   *   *

"You may have ordinary licentiousness, which you may see in all parts of London, in the public streets. That is a *vice* which is very difficult to deal with, except by the police; but that vice sometimes passes from the normal into an abnormal state, and the exaltation of the instincts becomes a disease, or mania."

Dr. Alexander Peddie, F.R.C.P., of Edinburgh, says :—

"The habit of drinking sometimes goes on in a slow, stealthy manner, as if from *vice* to disease, and binding in the last link the victim, as a slave of a passion from which he cannot free himself, struggle however hard he may. The disease therefore may be acquired, springing out of *vicious* courses."

Dr. J. Critchton Browne, of London, Lord Chancellor's visitor of lunatics, says :—

"Habitual drunkenness is a *vice*."

"Dipsomania is a disease."

"In drunkenness, the indulgence of the propensity is voluntary, and may be foregone, and in dipsomania it is not. The vice may become involuntary, and a disease."

"Dipsomaniacs are indifferent as to their relatives and friends, and their moral nature is degraded. With habitual drunkards that is not necessarily so. Burns, Collins, Poe and other poets have been addicted to drunkenness, and their later works have been as good as their earlier. In certain cases the drunkard becomes a dipsomaniac. If he is predisposed to nervous disease or insanity, he invariably becomes a dipsomaniac."

"When a common drunkard has irresistible craving, says he cannot resist it, his affections are weakened, his intellect fails, his will becomes weak, and he neglects his affairs and his family, he has crossed the line, from *vice to disease.*"

Dr. Geo. W. Mould, Superintendent of Royal Lunatic Asylum, Manchester, England, says :—

"Intemperance as a vice is the result of vicious and immoral habits. Intemperance as the result of a disease, is attributable to an impulse which the patient cannot control. In the other stage, he can be made to control it."

THE CRIME VIEW.

Crimes are undoubtedly committed by persons who are under the influence of strong drink; but it is important to discriminate between the criminal who drinks as he smokes, or indulges in any other vicious habit, and the criminal who drinks for the purpose of aiding him to commit crime. In the one case, drinking is but one of a series of evil practices, which may or may not be excessive and injurious. In the other case, liquor is imbibed for a purpose, and that purpose is to strengthen the daring or courage, and enable a person with criminal mind, to be bold or audacious enough to . proceed with his intent, and accomplish it. The one may be noted on the court docket, or on the prison roster, as intemperate or not, according to the judgment or habit of the justice, or the warden. The other takes care not to imbibe enough to induce profound intoxication, and thus betray himself, and yet he drinks enough to entitle him to be registered as a drunkard. He takes his draught of whisky, in just the quantity needed to harden his conscience for the time, and nerve his arm for the intended deed. He is the possessor of a criminal mind, and is bent on a criminal act. He is criminal first, and drinker afterward. On the court and prison records he has no right to appear as a drunkard. Possibly the least of all his vices, is his drinking habit, if indeed it reaches the measure of a habit. He procures his liquor for the same reason, and with the same deliberation, that he purchases his weapon. Indeed, his cup is one of his weapons. He avoids inebriation that he may succeed in his purpose. Among professional criminals, such men are not few. They are too shrewd to become intoxicated. Let me furnish illustrative cases, which I take from my own journal.

No. 78. Aged 29. Clerk. Drinks to excess occasionally. Latterly, the occasions have been so frequent that he has lost his situation. Upon further acquaintance with this youth I found him honorable and honest, when sober, an excellent clerk, obliging, and given to no other habitual vice, than excessive smoking. He is sensitive, and has recently become suspicious to a degree that makes intercourse and conversation with him a very delicate matter. At times, he was overtaken with an impulse to commit an act that was in violation of his conscience and moral sense, but which seemed to be irresistible. The very conflict with himself and his temptation, aggravated his nervousness, and he became willful, obstinate, profane, and restless to a degree that was irrepressible. In this stage of extreme irritability, he would resort to whisky in great moderation. Unlike the dipsomaniac, who drinks without limit, and without thought, he drank with great caution, taking a little at a time, with short intervals between. As the circulation

3

began to create a glow throughout his whole capillary
system, and his extreme nervousness began to yield to a
state of comparative calm, the period for deliberation
was reached, and keeping himself at this level, by
repeated draughts of liquor, at suitable intervals, he was
enabled to plan and execute. His offence was always
the same, and after it was done, he suffered remorse,
and sorrow, and till the next overpowering impulse
possessed him, he was prudent, sober, and correct.
This young man afterward settled in business, and
became a useful citizen. His friends consider him
a "reformed drunkard," and he is willing to accept
the title. He is, however, a reformed criminal, if the
propensity to crime is in subjection; but he was never
an inebriate, in its actual physiological sense.

No. 180. Age 41. An agent for a large mercantile
firm, who, with a clear head and steady hand, executed
a forgery, and then deliberately got drunk, to partially
obscure from his mind, thoughts of his deed, but more
especially to furnish his friends with a plea for committing him to an inebriate asylum, the officers of which
were unconsciously instrumental, for the time, in aiding
a criminal to escape the just sentence of the law. I am
not aware that this man was ever intoxicated afterward,
while previously to this time, he bore a reputation for
sobriety. He was not an inebriate, but a criminal, and
yet the fact of being sheltered for a short time within
the walls of a Sanitarium, gave him the opportunity to
pass, on his discharge, for a reclaimed victim of the
bowl, which he preferred to the shame of being a forger.

Another case that came under my observation, but not under my care, was a convict in a State Penitentiary for the third time, for manslaughter. I shall narrate no more of his remarkable history, than will serve to illustrate the topic in hand. Notwithstanding his homicidal tendency, which seemed to be inherited, he acknowledged himself to be a coward, and it was always with much fear, that the impulse to kill was associated. Instead of the daring, and even rashness of some homicides, he trembled with terror as the impulse to destroy life, seized and possessed him. The conflict between the impulse, and the timidity and dread, which were almost simultaneous in their approach, made him nervous, irritable and angry. Like Number 180, under these conditions, he resorted to the liquor in such carefully graduated quantities, as he imagined would secure care and deliberation in the prosecution of his purpose. His purpose was to select a victim whom he could manage with ease, always keeping himself in the attitude of self-defence, that he might evade the extreme penalty of the law for murder. He could, while his own anger and irritability were under control, excite his antagonist to threats, or attempted assault, during which period he would calmly and surely inflict the fatal wound, under the pretence of saving his own life. This sort of proceeding had been practiced with success three different times, on which account he had spent most of his adult life in prison, and before his present term expires, he will probably die in his cell. He described to me with evident clearness, and certainly with considerable self-satisfaction, the details of his proceedings, and manifested no evidence of remorse on account of his guilt. The cause of his crime on the prison docket was "intemperance," and he was willing to accept this record as true, because it was written, for he did not appreciate the enormity of the crime, nor the guilt of a criminal. He should not have been so registered. He is not a drunkard, but a murderer; the criminal intent was in his mind; the objects of his assaults were selected, and plans laid to decoy and irritate them, before he drank the whisky to aid his brutal instinct, and nerve him for the fulfillment of his diabolical purpose. He represents a class, and I doubt not, if a careful analysis was made of the character and habits of convicts now in confinement, the discovery would be made, that many whose crime-cause is stated to be intemperance, would be found to be like the one just stated—temporary drunkards for a criminal purpose.

Inebriates proper are not inclined to criminal acts. Dr. Arnold, of Baltimore, speaks of them thus:

" Inebriates do not form that class of people who plan and carry out schemes of villainy and corruption, in high and low places; nor are they usually found on the list of professional criminals who figure in our courts of law. Besides, it is notorious how often criminals try to mitigate the heinousness of their offences by attributing them to the effects of alcohol."

It should be borne in mind also, that the very habit of intoxication, disqualifies persons from committing some crimes. The habitual and excessive use of intoxicants, promotes timidity, incautiousness and inefficiency, and failure is the almost invariable result of attempts to commit certain kinds of crime, by those who indulge in intoxicants. An expert was some time since employed to search the records of crime in a neighboring State, with the view of ascertaining from official sources, the number of persons convicted of murder during the past hundred years, with the causes, penalties, etc., etc. After a careful and painstaking examination of court and prison records, it was reported, that less than three per centum of such crimes could be traced to the use of intoxicating liquors. Upon this disclosure being made, it was repeated to a prominent temperance advocate of the same State, who confirmed its accuracy, by saying that he had caused a similar investigation to be made, with the same result, but added that he hesitated to make it public, because it would deprive advocates of temperance of a cogent argument in behalf of the cause. Pursuing the same line of inquiry from time to time, it fell in my way to ask a very worthy Chaplain of a Penitentiary, how many of the several hundreds of convicts under his care, could connect their crimes with the use of intoxicating drinks. His reply was, that from direct personal knowledge of the history of each prisoner, he believed they were all guilty of vices, such as gambling, profanity, falsifying, tobacco chewing, smoking to excess, and lewdness, etc.; but to which of these vices their particular crime was to be attributed he could not tell, but that it would be about as easy and fair to trace it to one as to another; and he added, "Those whose crimes are the direct result of intemperance are very few. *I do not know of one.*"

It would be more philosophical to go behind and beyond them, to the source from which they all spring, namely, a depraved physical and moral nature. Being children all, of the same stock, their conduct and behavior originated in one common source, and it takes either line that is indicated, in accordance with the direction of certain physical tendencies.

The Hon. Richard Vaux, of Philadelphia, distinguished as a penologist of rare powers, and opportunities for observation, writes me as follows :

"I do not consider intemperance, as it is called, —inebriety,—the use of intoxicants, as a crime-cause. If this were so, all inebriates would be criminals. Now, the fact is, that criminals are made so by other causes; and they, like the rest of mankind, use intoxicants or *do not* use them. It is now forty years since I have been an Inspector of the Eastern State Penitentiary in this city, and I have no hesitation in saying that intemperance—the use of intoxicants habitually, or to excess—is not a crime-cause. I think it can be said, that about one-half of those convicted of crime are total abstainers. Of the four hundred and thirty-three (433) prisoners received into our Penitentiary in 1881, but twenty-six (26) were intemperate. Mr. Cassiday, our warden, who has been in the service of this prison for twenty years, gives his experience in support of these views of crime-cause. I know it is a sort of fashion to talk about our prisons filled with the victims of intemperance, but the figures do not support this general and sweeping assertion."

In confirmation of the same views, I am furnished with the following, from the accomplished General G. Mott, late keeper of the New Jersey State Prison, at Trenton :—

"I am decidedly of the opinion, that our Penitentiaries are not filled with those who trace their crimes to intemperance; that class fill our common jails, lock-ups, and Houses of Correction. A person sent to a Penitentiary, no matter for how short a time, for a violation of the law, perhaps committed in the heat of passion, and while under the influence of liquor, is branded a criminal; thinks society has injured him, and when he gets out, may join the criminal class, as he says, 'to get square;' but he must keep sober if he expects to get in with the expert. The majority of criminals who fill our Penitentiary are primarily of a criminal mind, born so, and brought up to prey upon the general community; but they are not habitual drunkards, nor do they associate with that class. Not so themselves, because, to be an expert, they must keep their heads cool, and their wits about them; and their associates must do the same, as they know there is no dependence on a drunken man; for, when in that condition, he may let something drop that, perhaps, will lead to the failure of their plans and the probable detection of the principals."

The following is also contributed from the Maryland State Penitentiary.

"Out of five hundred and thirty-four (534) convicts in November, 1881, there were strictly temperate one hundred and seventeen (117); moderate drinkers, two hundred and forty-two (242); occasionally intemperate, one hundred and seventy-one (171), and habitually intemperate, four (4)."

From Mr. John C. Salter, the successful Warden of the State Penitentiary at Chester, Illinois, I learn the following:—

"The popular sentiment seems to be that a criminal must necessarily have been a drunkard. That this class are frequenters of saloons, and are more or less slaves to appetite for strong drink, as they are to other vices, cannot be denied. The large proportion of criminals, such as burglars, forgers, counterfeiters, need clear brains, steady nerves, and quick perceptions, to successfully carry out their plans, which would be impossible under the influence of intoxicating drink. I am more and more convinced that the causes of crime go away back in the history of the criminal, even outside of his own life, coming down from generation to generation, visiting the iniquities of the fathers upon the children. Lack of home influence, throwing boys and girls of tender age out upon the charity of the world; lack of the discipline of education, the haste to get rich, and the false standard of greatness; are causes that have done much toward filling our jails and penitentiaries with those who, under more favoring winds, would have found shelter in a friendlier harbor."

The apparent discrepancy between the commonly accepted belief, that at least two-thirds of all the crimes are due to intemperance, and the actual facts, as derived from institution statistics, may perhaps be accounted for thus: The offences for which persons are sent to houses of correction, county jails, and lock-ups, are largely attributed to strong drink as an *exciting* cause, while the more grave offences are punished by commitment to penitentiaries. Also, the commitments to the common places of detention, are counted over and over again, and the evil is made to appear, as we shall presently see, much more formidable than the facts really justify. Vagrancy is an offence that does not find its way to penitentiaries, and yet it occupies a conspicuous place on the common jail records. Vagrancy is often associated with drunkenness, but not always as cause and effect. Pauperism and vagrancy are usually associated with a low and depraved physical and moral constitution. In many cases, the tendency is to despondency, and despondency is frequently an exciting cause of intemperance. If, therefore, vagrancy is counted as crime, and every vagrant who drinks is counted as intemperate, it can be readily seen how so large a percentage is given to intemperance as a crime cause. So, if intoxication is counted a crime, and, to use the police nomenclature, if "drunk and disorderly" is a title attached to every commitment for intemperance and vagrancy, the showing in that direction must necessarily be exaggerated. And yet it is just about in this careless manner that the police records are frequently kept. A scientific nomenclature is unknown to the law, while the docket of a police justice, cannot be more than a transcript of the justice's own ideal of what is, and what is not crime, or disease. Crime usually has its source in the mental or moral

constitution. The desire for alcoholic beverages is
generally a physical desire, an animal lust, and has but
a distant, if any, relation to what is recognized as the
moral character. Poison acts primarily upon the blood
and upon the nerve centres, while the moral aspect
of the case is remote and incidental. The American
Association for the Cure of Inebriates expresses this
thought forcibly in the following resolution, which was
adopted at their Annual Meeting at New York city, in
1871.

" *Resolved,* That the effect of poison on the blood and
nervous system, and the reflex action of this morbific
agent upon the whole physical structure, is the same in
the virtuous as in the vicious, and that antecedent or
subsequent moral conditions are incidental to the main
fact of disease."

As drinking is the outcome of physical states, so also
social conditions contribute their share towards the
same result. Bad food, unventilated dwellings, low ·
companions; whatever contributes to lower the tone of
bodily vigor, tends also to drunkenness; and yet, in
arranging a table of causes, these important factors are
but seldom, if ever, noticed. It is a question, the solution of which is now approaching its crisis, whether
vagrancy is not as much an exciting cause of drunkenness, as intemperance is of vagrancy. Certain it is,
however, that it is as unphilosophical as it is unjust, to
formulate a law or rule upon which to base conclusions,
without a rigid analysis, not only of criminal records,
but of the life, previous character and tendencies of

those who commit crime. To do this under the present system of keeping such records is an exceedingly difficult task. To register commitments, and not persons, is a most misleading method, to say the least of it, and until it is abandoned, and a system based upon actual fact, and individual knowledge, is recognized as proper, and adopted, we shall, perhaps, never make much progress. I know of a single man being committed to an asylum, and discharged thirteen times, and registered each time as cured of insanity, thus being made to represent on the record, thirteen distinct cases of insanity, and thirteen cures. The student of crime-cause and prison discipline, must frequently notice this method, and be impressed with its unfairness.

The criminal history of Great Britain furnishes some striking facts in this connection.

A noted " repeater," or " prison-bird," as such persons are sometimes called, was the notorious Margaret Mitchell, of Scotland, who was subjected to forty-two commitments, mostly for drunkenness and disorderly conduct, being two hundred nights in police stations, and seven hundred and seventy-eight days in Glasgow Prison, and in various jails, shelters, refuges, hospitals, etc., for short terms, in each of which she was, of course, registered as a distinct and different individual, and was made, under her forty-two separate commitments, to represent an equal number of cases, being variously recorded with the appended offence, as " drunk and disorderly," "indecent conduct," "annoying her husband on the street," "annoying her husband at his place of business," "outrageous conduct," "outrageous and profane language," etc., etc.

To continue extracts from the Parliamentary Blue Book, containing "minutes of evidence" before a Committee of the House of Commons, I find that William Smith, Governor of the Prisons at Ripon, reports as follows :—

"I have one case of drunkenness; the case of a woman who has been in Wakefield jail seventeen times, in Leeds jail eleven times, in Northampton jail fifteen times, in Ripon jail fifteen times, all for being 'drunk' and 'drunk and disorderly.'"

This woman was committed fifty-eight times, and if she did not represent fifty-eight separate individuals, she did count in each jail as a separate case, under a fresh commitment each time, and under a new alias, as Thompson, alias Fox, alias Connelly, etc., etc. Mr. Smith also reports the case of a man who was enrolled · in several prisons as Wolf, alias Blanche, alias Murray, alias Johnson, who, during a period of eight years, was sent to five different prisons twenty-two times, and in each case there was the offence of assault added to that of drunkenness. I might repeat a variety of illustrations of a similar character, from the same source, but these are sufficient to emphasize the fact that such records are not reliable, as indicating the existence of crime, as to its extent, or the number of its victims. With me it is no longer a question of doubt, whether intemperance, vagrancy and pauperism are not triplets in the vice and

crime family, having a common origin, being born of
one parentage, dwelling and thriving in a common
atmosphere, and capable of living separately or together,
as circumstances or accident may determine. The records
of crime and disease, and the observation of experienced
observers confirm and ratify the opinion. Certain it is
that phthisis, insanity and epilepsy, and possibly other
diseases, existing either together or separately in a
family, constitute a basis for the physical character of
offspring that is transmissible, and is transmitted as an
inheritance that is alike common to criminals, lunatics
and dipsomaniacs. In the 52d Annual Report of the
State Penitentiary for the Eastern District of Pennsylvania, I find a most suggestive and instructive table,
showing the relationship between crime and disease in
sixty-three convicts, which seems to confirm the above
statement. The table designates certain diseases that
stand related to crime with such uniformity as almost to
warrant the expectation of a similar coincidence of relation, if applied to the whole catalogue of crimes. In
these cases, larcenies, felonies, conspiracies, arsons and
murders seem to have substituted the diseases which
have manifested themselves in the preceding generation,
and to play an important part in the family history,
exhibiting the constant interchange between physical
and mental disorders.

For example, eight cases of manslaughter, and murder
in the second degree, twenty-six of larceny, and numerous other crimes in various members, and all, with
scarcely a variation, coming from families in which either

insanity, epilepsy or phthisis existed as the recognized family disorder, and in some of the families each of the three diseases did its share of destructive change, which was visited upon the next generation in the form of crime.

The following extracts from the testimony given before the House of Commons Committee on Habitual Drunkards, is offered to confirm the views of this chapter.

Dr. Arthur Mitchell, of Edinburgh, the distinguished Commissioner in Lunacy for Scotland, says :—

"There is a difference between ordinary criminals and the man under restraint for habitual drunkenness." * * * "I think you would do him (the drunkard) no good, if you did not in some measure treat him as a person laboring under disease, and if you place him in an ordinary prison you must treat him as an ordinary prisoner." * * * "The prison is not the proper place for the treatment of drunkards."

Mr. William White, Surgeon and Coroner for Dublin, Ireland, says :—

"I think that men with delirium tremens very rarely commit homicide. They are cowardly, and seldom have courage to attack any one."

Mr. H. Webster, Governor of Prisons at Kingston-upon-Hull, England, says :—

"The expert thief is always a much sharper class of prisoner (than the inebriate). He is quite a different type, I do not think that they are lower than ordinary criminals, excepting after they have given way to drink, and then their minds become impaired."

"Drunkards are altogether unfit for prison ; they are not the men to be associated with ordinary prisoners. I would never commit a drunkard to a prison at all."

"Small offences, petty larcenies and that sort of offence, are almost invariably the outcome of drunkenness" * * "not great offences, but petty offences." * * "Almost universally, in cases of robbery, the prosecutors have been drunk at the time."

"Those sort of cases, almost invariably arise from drunkenness on the other side."

Dr. Charles Robert Bree, a practicing physician of Colchester, England, says :—

"Drunkenness is not a crime like that committed by those with whom they associate in jail ; and they get a good deal worse by being sent to jail."

Mr. William Smith, Governor of Prisons, at Ripon, Ireland, says :—

"Those who are habitual drunkards are not really thieves: there is a distinction." "A thief is one who undoubtedly takes to drink to a certain extent, but he always takes care about getting drunk, because if he gets too drunk, he cannot commit the felony."

"One woman was committed for drunkenness fifty-eight times, but without having committed a felony once." * * "Out of eighty prisoners, eight to ten were drunkards."

INEBRIETY A DISEASE.

In entering upon the consideration of the aspects of disease, it is important to discriminate between the disease proper, and its morbid anatomy, in order to obtain a clear conception of its character. The unprofessional reader will be aided in his ability to comprehend this difference, by noticing the following definitions: In any given case of disease, the enquiry naturally embraces the causes, the symptoms, and the morbid effects produced upon any of the organs or tissues of the body. The causes and history of a disease are spoken of as its *etiology*. The symptoms, as its *symptomatology*, and its morbid effects upon the organs, as its *morbid anatomy*. Neither of these, however, will furnish any correct notion of pathology, an idea of which we shall obtain as we consider causes and symptoms, and trace their connection with a primary state of the system, in which we recognize a special tendency to particular forms of disease, known as a constitutional diathesis or predisposition. With the symptoms and effects of alcoholic intoxication most persons are supposed to be familiar in a general way, and yet there is a degree of ignorance of causes that is not in keeping with the general intelligence of the people on most subjects.

Not far from a century ago, Dr. Rush, of Philadelphia, attributed drunkenness to a "morbid state of the will." Esquirol, a distinguished French author and teacher, said: "There are cases in which drunkenness is the effect of accidental disturbance of the physical and moral sensibility, which no longer leaves to man liberty of action." Dr. James Critchton Browne, recently of the West Riding Asylum, England, says: "Dipsomania consists of an irresistible craving for alcoholic stimulants." * * * "Sometimes the craving becomes altogether uncontrollable." * * * "I recollect the case of a gentleman, perfectly sober, who had dipsomania, which was attributable to taking a draught of water on a hot summer's day, which caused fainting, and was succeeded by an entire *change of character.* Dipsomaniacs have no energy or fixity of purpose or strength of volition."

Dr. Alexander Peddie, of the Royal College of Physicians, Edinburgh, Scotland, says: "Intemperance as the *result* of a disease, is attributable to an impulse which the patient cannot control; the habitual drunkard, or dipsomaniac, or by whatever name he may be called, is destitute of control over himself. In that condition, no considerations, temporal or spiritual, have the slightest effect in checking the drunkard's progress, if ways and means, foul or fair, can be found to gratify the desire for alcoholic stimulants."

If we mark the chief features of this disorder as presented by the authorities just quoted, a fair estimate may be formed of the pathology, or true nature of the disease.

"A morbid state of the will," (Rush). "Which no longer leaves to man liberty of action," (Esquirol). "An irresistible craving," "altogether uncontrollable," sudden shock, causing "fainting and an entire change of

4

character," (Browne). "An impulse which the patient cannot control," (Peddie) etc. I might continue quotations from divers authors, of the same character, but prefer to illustrate the facts stated, by cases from my own clinical record, as follows:—

CASE 98. An educated gentleman, and an instructor of others. Trained religiously, and free from common vices. Domestic in his tastes and habits. Studious, and inclined at times to solitariness, when he usually resorted to his study, and desired to be left alone. This was generally the prodromic sign of an impending debauch. With locked door, and introspective mood, he would sit by the hour, meditating and struggling with endeavor to crowd out, and banish from his thought the propensity to leave his home, and drink. It came upon him unbidden, lingered with him, and at last became imperious and commanding. The resistance of his conscience and his convictions, seemed to intensify the force of the impulse, and, of course, to aggravate his disquietude, and increase the desire to indulge in what he knew would cause him to forget his troubles. Parental remonstrance, the attempted assertion of his own manhood, the resistance of his moral nature, were of no avail. In addition to these, the prospect immediately before him, of the low saloon, the boisterous company within, the emptied purse, the arrest, the lock-up, and the resulting shame and remorse, all passed like a panorama before him, but he heeded nothing, till his debauch was accomplished, terminating, as it always did, in protracted oblivion. He invariably attributed his defection

to indescribable sensations at the epigastrium. Placing
his hand over the region, he would cry out. "Stomach!
oh, stomach! why do you torment me?" And speaking of his other sensations, described them as accompanied by a sense of darkness and sadness, enveloping
him, as it were in a cloud, by which he was borne
almost unconsciously from his home, to the haunts that
he frequented. What he did after the first draught,
which was invariably full and strong, he knew not. I
do not mean that he was in a drunken sleep all the time,
for he was not. On the contrary, he was active, busy,
and restless, but he was led by a force that he could not
master. He seemed, in movement and utterance, to be
purely automatic; an invisible force guided his steps,
and at last, held him in the bondage of a complete narcotism, that lasted several hours, and was the invariable
culmination of his attacks. When the chain was broken,
and the emancipation was complete, he could recall
nothing that happened after his first drink. Months
would sometimes intervene before another attack. It
was always announced, however, by the same prodromic
period.

His stomach disorder was concealed for a long time,
because it was associated in his mind with debauchery
and shame, and a desire for seclusion and loneliness.
It might be called the second stage of his attack, the
third being a period of active, restless and boisterous
intoxication, terminating in a long and profound sleep.
These were his uniform symptoms. The climacteric
period is now passed in his case, and he is a sober, useful,

Christian citizen. Review the points here stated, a moment; a thoughtful, educated gentleman, no training by a previous habit of moderate drinking, no influence from association with drinking companions, no temptations from without, but a deep-seated nervous disorder, gnawing at the vital parts, and demanding relief, but aggravated, doubtless, by a family taint of insanity! These are the points.

Dr. Arthur Mitchell, Commissioner of Lunacy for Scotland, says: that in some cases "frequent habitual drinking precedes this state, *but that it is not necessary; it may sometimes appear without previous habits of drinking*, as the result of cerebral injury; the result, for instance, of fever, of hemorrhage, of mental shock, of the commotion in the system which attends the establishment of puberty, or the arrival of the climacteric period."

Dr. Geo. Burr, of Binghamton, New York, says: "It is the propensity or desire to indulge in the use of ardent spirits, and not the habit of drinking to excess, or drunkenness, and its subsequent effects, that is to be regarded as the morbid condition. A broad distinction must be made between the two; the latter is but the development of the former. The propensity, when under the influence of the exciting causes, arouses the appetite, overcomes the will, blunts the moral sensibilities, and makes everything else subservient to its demands. The habit is the natural sequence of the growth and development of the propensity. The one bears the same relation to the other that the eruption of smallpox bears to the contagion of that disease; or the several stages of an intermittent fever, to the poison of malaria. The delusions of the insane are but the morbid phenomena of minds disordered ; so, the love of the bowl, and the self-destructive acts of the inebriate, are likewise the manifestations of a condition of the organism, that may well be regarded as diseased." Here we have a pathological state referred to, that is often described by those who possess it, as a state of unrest. Dr. Arthur Mitchell, already quoted, describes it as "an ungovernable and remitting craving for drink, which has no reference to anything external ; it comes from something within."

An ungovernable craving ! that is the pathological state, whether it originates in the nerves of the stomach, in the brain, or elsewhere. Whether it is the result of inheritance, of imprudence, or of accident.

There is another view, that traces its pathology to the alimentary system, that is worthy of consideration, and I offer the following extract from a lecture I delivered before the Medical and Chirurgical Faculty of Maryland, in 1874,* in which the following language was used : " Our appetites differ widely as to the ordinary articles of diet; what is delicious to the taste of one, is offensive to the taste of another. Especially is this manifested in eccentric diversity, in certain forms of disease, as in hysteria and chlorosis. Lime, chalk, clay, etc., are, as we all know, eaten with relish in such cases ; and how often do we see children, eat plaster from the

* Vide " Transactions" for that year.

wall, slate pencils, etc., not, certainly, because they satisfy
the gustatory sense, but because there is a constitutional
demand for them, to which even the common instincts
of childhood answer, without the intervention of any
theory or knowledge on the subject. Such cases evidence a disordered alimentation which the physician
proceeds to correct, and which is amenable to his corrective processes. We are also familiar with analogous
eccentric demands of appetite or longing, in pregnancy,
and in cases of convalescence from fevers; the strongest
and apparently most indigestible articles of food are
frequently required, and I suppose, generally allowed,
in such conditions, with no apparent disadvantage." I
therefore place the excessive demand for alcoholic stimulants, in some cases, among this class of diseases.

It is quite probable that in the organic changes which
take place in the processes of life, and which we call
waste, there are periods when the diathetic condition,
originally occult in the germinal existence, may become .
apparent, and manifest itself in appetites or passions
which were before not known. We see this fact exemplified in the growth of the individual. The infant has
no desire or capacity for the food of manhood; but with
dentition and early childhood, new appetites begin to
declare themselves, which are modified as life advances.
Waste may not always be supplied by the same material,
even in the same individual. Age, occupation, climate,
and habit of life must, to a considerable extent, regulate
the quality and quantity of the alimentary substance;
and as there are conditions in disease which require

alcoholic stimulation, so there may be changes in the structure, from natural or accidental causes, which seem to demand their use, and in which the demand is sometimes imperious. I may in this connection, also invite attention to that inordinate craving for food which is known in science as Bulimia or Polyphagia, in which condition there is an excessive demand for ordinary healthy food, in quantities far beyond the needs of the system. It is commonly called gluttony. It may be, like drinking, a sensuous habit merely, running on till it assumes the form of gastric disease; but it is more frequently a symptom of cerebral disorder. There are certain forms of mental derangement, in which it is a characteristic symptom, sometimes assuming the form of a voracious demand for unnatural food, as for raw or spoiled meat, candles, etc. I have seen it frequently manifested in idiocy, to a most revolting degree. It differs from the revived and invigorated appetite of convalescence from protracted fevers, and is one of the most offensive and disgusting types of animalism to which the human species can be reduced. Allusion has already been made to the chlorotic appetite of females; a morbid craving for particular things, and especially those that are innutritious, which Dr. Flint says is sometimes due to a " morbid uneasiness at the stomach." This is the very symptom so often referred to by intemperate persons, and the more I compare the pathological states of the different classes, the more striking does the similarity of condition appear. The stories of dirt eating that come to us from the plantations of the West India islands,

and of our own Southern States, intensify the fact and
emphasize its pathological importance. Dr. Duncan, of
Louisiana, believes this practice of dirt eating (Chtho-
nophagia) to be due in many cases to "a diseased
condition of the digestive organs."

Again we find one other feature common to all these
states, namely, a certain perversion of the moral sense,
leading to deception and falsehood, regarding their habit.

The intimate relation that exists between the moral
and intellectual tastes and propensities, and those that
are physical, and the retroactive influence of mind and
matter, lead to the thought, that while it is true, that
disordered alimentation in some cases is due primarily
to gastric disturbance, or to morbid impressions upon
the gastric nerve centre, it is also true that in other
cases, the abnormal sensations are due to the brain, or
to that portion of the brain which has to do with the
function of taste. As hallucinations and delusions are
evolved from that portion of the encephalon that is
concerned with the sense of hearing, and which in such
cases is so distorted by disease, as to create false sounds,
so may the desire for alcoholics, be evolved from the
convolutions that are concerned in the function of taste
and appetite, if from any cause that portion of the
cerebral mass is under morbific control.

Whether the disease, therefore, is one originating in
the brain, or in the stomach, or sometimes in one and
sometimes the other, is a question which naturally presents itself in this place. That it may originate in the
brain is quite true; that it may originate in the gastric

organ is equally true; that it may be idiopathic, or traumatic, is also true, and with but little difference in its symptomatology from either variety of causation. That it may exist in a form that may be called psychical inebriety, is also capable of demonstration. Though I believe such cases are rare, and because they are rare, or at any rate, are but seldom described, if, indeed, they are noticed, I shall now give the outlines of two cases, —a father and son, both of whom were victims of this unusual variety of the disease.

CASE 208. A young man possessing rare gifts of mind, an only son, in many respects the counterpart of his father. They were both professional men of ample means, and with but little to think of, except how best to enjoy life; and of course each had his own ideal of what constituted enjoyment. The father was an extremist in religion of the transcendental order, and seemed to dwell in an atmosphere that imparted to his inner sense, the most exquisite delights, and when not ranging in invisible spheres, and communing with unseen friends, he was intent on securing converts to his faith; and especially was he anxious to enlist the gifted mind of his son, in the same pursuits with himself. The son, on the other hand, could not adopt his father's ideal, though he was envious of his ecstatic flights, and determined to avail himself of the intoxicating and bewildering effects of ardent spirits, hoping thereby to arouse, if possible, similar ecstacies to those of his father's mental state. His judgment could not accept the religion of the father, though he thought he discovered that its realm was, to a

great degree, within the scope of a lively imagination,
and that by stimulating his own powers, he might
occupy the same field, and enjoy similar fellowships and
fancies.

Both parent and son were alike in temperament; the
bodily health of each was good, and on more than one
occasion, both in my presence, and in the presence of
each other, were earnest and sincere in argument and
appeal, to convince me that the other was insane. The
son conceived the father to be a monomaniac on the subject of religion; and the father believed the son to be
insane, because, not accepting the dogmas of his trans-
cendentalism, he obtained enjoyment from the bowl. The
brain of one was disturbed by a faith which inspired
his conduct to a degree, and in a manner, to warrant his
being classed with those who—

"Are drunk, but not with wine,"
and who—

"Stagger, but not with strong drink."

The brain of the other, was so far athwart its balance,
as to believe that he could substitute the intoxicants
for a religious faith, and draw from their inspiration
similar delights and enjoyments. By unreasonable
methods, both sought to realize what they could not
possess in a normal state, or could not obtain by reasonable means. The recompense to each, was in harmony
with his tastes, and with the means employed to indulge
them.

These men occupied the border land between sanity
and insanity, for a season. They kept pace with each

other in concurrent lines, during several years, each following his own course to its end. Occupying separate homes was among the early signs of domestic dissolution, and the sequel of the son's career was a permanent lesion of the brain, requiring a care-taker for the remainder of his life.

The natural outgrowth of his father's vagaries was a gradual but continuous loss of mental poise, and a corresponding diminution of worldly fortune. Both of them, from a common impulse, that was purely psychical, sought happiness through channels, that were alike familiar and congenial with their tastes, but leading to one and the same result. The son reveled in an artificial atmosphere, the product of alcoholic intoxication. The father delighted in a rapturous communion with a counterfeit world, which was brought within his reach from beyond our own sphere, not by the poison of alcohol, but by the toxic wand of a bewildered imagination. The brain was intoxicated in both cases, and yet neither was an inebriate. The father exhibited psychical, and the son physical symptoms of intoxication.

Thus far our investigations have led us to recognize the longing desire for intoxication, and the ungovernable craving, which has no reference to anything external, and not the act of drinking, to be the disease, the cause for which may be hereditary, or may be created by the conditions and conduct, which interweave themselves into daily life.

· There are certain peculiarities, however, which are displayed with such remarkable uniformity in the career of some inebriates, as to warrant their classification into separate groups, and I shall now offer a few examples in proof of this statement.

SOLITARY, MIDNIGHT INEBRIATES.

CASE No. 59. A clergyman, aged 46. Drinking habitually for six years. Nervous asthenia, melancholy, with hallucinations. This is a representative case. A gentleman in the prime of life. His profession an accepted warrant for uprightness and probity. A victim of neurasthenia,—nerve exhaustion. Worn and debilitated by pulpit and pastoral labors, he had expended more force than he could afford to spend. He sought to replenish his wasted energy in the retirement of his own study. His calling forbade the public use of intoxicants. He never entered a saloon, and if wine was offered at social gatherings, he forced himself to refuse it. He was an ardent student, and in the silence of the long night hours, while his family slept, he was to be found in his lonely study. There, with his bottle and his books, his wearied nerves contested with his conscience, for the right to forget his cares, and invite his slumbers by the free use of the bottle. At first the conflict was severe; there was no localized distress, and yet every nerve tendril seemed to throb and yearn for repose. A sort of moral mirage enveloped him, and his thoughts lost themselves in the mist. Then he emptied the bottle, and the sleep of intoxication put an end to thought.

The following day brought with it a parched mouth, a dry tongue, headache, irresolute purpose, and remorse. Is it any wonder that seasons of melancholy sometimes come, to bow the spirit with forebodings of coming evil, and that countless spectres of unreal things should float in the atmosphere of such a mind? I can scarcely conceive of a state more miserable. Secret intoxication, lasting only for a night, to be followed in the morning, by the semblance of a normal life, visiting and counseling, holding meetings, and conducting public worship, and all the while, behind the invisible mask, the memory of the closet debauch, stinging the spirit, with no other relief than in its repetition. Carrying within him the image of the scene he cannot forget, and yet wishing the hours to hasten on, that he may cast aside the mask, and behind the locked door, the bottle, the drink, and the narcotism, may rest him again for another day.

This is no fancy sketch. Many an eye may rest on this page that has witnessed similar struggles, and fallen into the same forgetfulness. There are many such men, men of learning, integrity and piety. They are solitary, night inebriates, who avoid alike the glittering saloon and the giddy circles of social revelry, but who almost nightly retire to their secluded Gethsemanes, where they may commune with the profoundest needs of their nature, and know not how else to supply and satisfy them. The world does not know their frailty, much less, the means they take for relief. To this class belong orators, and literary men, men of genius. Such cases are apt to terminate in some form of chronic alcoholism, the evidences of which become apparent in attacks of local or general paralysis, or some similar neurotic or brain disorder. Dr. David Skae, an eminent alienist of Great Britain, and physician to the Royal Edinburg Asylum, says: that "there is the regular drunkard who keeps sober during the day and gets drunk at night, and attends to his business regularly during the day. Such men carry on for many years, without injury to themselves or others. I have known one case, where a gentleman was carried to bed drunk every night for *fifty years*, and yet he made a large fortune, and was in the market every morning, attending to his business." * * * "Many of them, I know, live to a great age; I have known men upwards of eighty, who have been drunkards of that kind."

The craving for liquor in these cases is different from the craving in some other forms of the disease. There are no long intervals of sobriety which are suddenly interrupted by an unexpected outbreak of a passion for drink. The craving takes the form of habit, more like that for a night meal after a long interval of privation of food, but it is nevertheless positive in character, and regular as to time of occurrence. That they sometimes terminate in paralysis, is illustrated by the following:—

CASE No. 69. Aged 73. Retired gentleman. Has been drinking regularly for forty years. Loss of memory, partial paralysis, etc. Such is the abstract of the entry made on his coming under my care. Forty years' drinking, but not often what is called drunk. A solitary night drinker. Attended to business during the day, took part in public affairs, was reported to be a man of experience and judgment, and well esteemed as a good citizen. One feature of his drinking was that his potations were not always excessive, and failing to sleep soundly, he would rise from his bed, appear on the street during the night, call at the houses of friends, and at accustomed business places, and finding them closed, would return, without confusion or embarrassment, to his home, in a sort of semi-conscious, bewildered state, go to his bed, and be refreshed by sleep. With the return of the morning, however, he failed to recall any of the transactions of the previous night. Following this mode of life, in a short time, partial paralysis succeeded the loss of memory, and his motorial functions being on this account interfered with, the night walks were discontinued, and exercise of any kind becoming difficult, he slowly yielded to the progress of his paretic disorder, and "death from dementia" was the final verdict of his medical attendant.

Solitary or secret drinkers are to be found in almost every community. Some of them will never come to the surface. Perhaps a majority of them are able to conceal their habit, and it may never be known. They are men generally of mark and of work—intellectual work—men of genius often, men whose habits, aside from the use of intoxicants, incline them to late hours, to study, and to seclusion. And we know not how many effusions of poetic genius, or how many discoveries in the realm of science or philosophy, have been evolved from amid the fumes of the bowl. For it is certainly true, that the impression made upon the human constitution by such beverages, may both hinder and quicken brain force, so, that under its influence a certain degree of brilliancy may be imparted to the mental operations of some persons, while in others, the consequence of its use may be confusion and incoherency. In these secret night-drinkers, the craving which constitutes the morbid state does not manifest itself, as it does in persons of different temperament and habit. It is a neurasthenia, but it is associated with the idea of privacy, secresy, concealment. Drinking is not spoken of. It is regarded as a practice not to be made known. For such a person to visit a public saloon and drink, would be as unusual and repulsive to the moral sense, as in the case of a temperance advocate. It does however, happen that there may be a divergence of this impulse into other lines. Like a current of electricity that follows a single wire to a point where the wire is connected with several others, the current will separate into many, and flow in each. So a certain constitution may, up to a certain point, be direct in its leadings, and then, from age, difference in condition or circumstance, exhibit other tastes and follow other pursuits, in which case the craving will be modified as to time of appearance, and the character of its demands. A most interesting example of this sort may be exhibited in the following :—

CASE No. 205. Age 26. Student. Habitual use of alcoholics, opium and hasheesh. I found this young man cultured, refined in taste and tendencies when sober, fond of music, pictures, books, and especially of the ancient classic authors, but when intoxicated, brutish, violent and dangerous. He speaks thus of himself: "I was born with a peculiar temperament; while in my cradle, was peevish, cross and troublesome, and have been a trouble to others ever since." As a youth, he was timid, retired, almost to solitariness. He says again: "As the 'teens gathered upon me I struggled with my timid nature, and sought the society of virtue and love. I was a big infant, timid, suspicious and fretful, always wanting to realize something that was not within my reach, but in the society in which I mingled I found not the solace that my nature yearned for." At college he was an isolated student. The sports which others enjoyed, he could not participate in; he learned to drink at Princeton, and was dismissed. He tried again at Cambridge, but with the same result. At last he became his own tutor, in the silence and seclusion of his own rural home, where, too, he became a solitary inebriate. While under my observation this youth presented for my study the most extraordinary combination of gifts and qualities. He was a specimen of enigmatized humanity, such as I had not before seen. Problems, riddles, concealments, obscurities, wit, culture, cheerfulness and sadness, obstinacy, refinement and rudeness, were all joined and jointed in a grouping that was to me most novel, and at the same time the most engaging. A family history was behind it all, covering generations, in which eccentricities were a marked feature. To any one who was patient enough to study him, and to whom he would give his confidence, he kaleidoscoped himself

5

hour by hour, presenting under various circumstances a most instructive psychological study. It is not surprising that such an one should invade the whole field of artificial stimuli in search of satisfying potions, and he tried most of them. Whisky at first aroused the most dangerous qualities of his nature, and under its influence, the quiet student and patient artist, was transformed into an excited madman, finally breaking away from his midnight solitude and roving abroad, unconsciously displaying himself to his neighbors. Recovering, and being told what had happened during his debauch, a turn was taken suddenly, and a new character appeared. To avoid alarming those about him, and disgracing himself in his own eyes, he determined to resort to opium, and thus he soliloquized: "Opium may bring a solace to a spirit wounded by the unremembered story of outrage and evil. It may bring rest to tired muscles and nerves. It may soften the passions, and subdue the misguided will. It certainly will bring repose and dreaminess to a jaded and wounded spirit. But then the effect may soon pass away, and leave the unhealed system to itself again. Thus doubting and fearing, the experiment was made, and for a time persisted in, but did not prove satisfactory, and again he soliloquized as follows: Opium does not satisfy. I am the same restless and untamed nature as before. It does not intensify any emotion, or thought, or desire. It simply soothes, dreams, and vanishes. It lacks positiveness and certainty. It will not do.

Now hasheesh is to be tried. It was taken freely, for

he could not bear long intervals of sobriety. The private drinking dispensation had passed with him. He was out on the high road of a reckless and lawless life, and yet every time he reviewed the story that was told him of his conduct during a debauch, the sensitive and refined element of his nature would assert itself, and an impulsive effort to change his course would be made. Now, he was in the fairy land of the hemp-eater. A singular sense of strangeness seemed to possess him and every object that came before him. Everything was tinged with novelty. Voices were strange. Distances were magnified. Sounds were unreal. Melody arose from his piano as a vapor that vanished in the distance, and lingered only like a far off murmur. Nothing was satisfactory, and he awoke from his half slumber, bewildered and distressed. Another sober interval, and all that he could learn from others, and all that he recalled himself of the Opium and Hasheesh experiences were studied and carefully analyzed. One was soothing, swaying dreaminess. The other was a puzzling, awkward sense of actual things, being unreal. Everything was perverted and no repose came. Thus were comparisons made with the alcoholic intoxication, but none of them compensated the disturbed nature with a satisfactory return. Nothing paid but the dead-drunk condition which alcohol creates. With all its unconscious profanity and brutality, it was the most satisfactory. It brought profound sleep. Absolute forgetfulness of self, and everybody and everything. It satisfied to the full the ungovernable physical craving, and terminated the paroxysm with longer intervals of a sober selfhood, than could be obtained by using any other intoxicant.

Recovery was the sequel of this case. He now lives a quiet, retired life, married, devoted to study, and surrounded with the refining influences of culture and taste.

Another form of intoxication whose symptomatology is unusual, is accompanied with a propensity to thieving, to drinking on certain days, and to commit violence, and frequently violence of a specific kind.

INEBRIATES WITH TENDENCIES TO SPECIAL VICES, AND TO SPECIFIC FORMS OF VIOLENCE.

Some are exceedingly economical during a carouse, and so manage to satisfy their desire for drink through. the pockets of friends or fellow-drinkers, and only drink when they are invited. On the other hand, a lavish wastefulness characterizes some, and they are ever ready to treat, not only friends, but any who may be near at hand. These peculiarities are not accidental, but uniform in the individuals possessing them. Some drink only on Saturdays, and oftener after sundown on that day, sleep off the fit, go to church the next day, and are sober the remainder of the week. Some have particular days on which they become debauched, and do not indulge at other times. Public holidays, private anniversaries, as of marriage, or the birth of a child, are occasions that are thus celebrated. I have known the proprietor of an extensive mercantile house, who, for eleven months and two weeks of each year, applied himself closely to business, in which he was successful, who, on the same day of the same month in each year, joined himself to a man of low estate, with whom he could not be on terms of social freedom when at home and sober, and occupied two weeks on a fishing excursion. Before the arrival of the appointed day he arranged his business for a fortnight's absence, drew checks to meet the wants of his home and his store, made appointments for the day of his return, and equipped himself to meet his companion, who was to serve as his guide and care-taker. With a boat on the bay, or river, a tent for the shore, and an abundant supply of "creature comforts," they commenced at the appointed time. Fishing by daylight, and tenting at night were to be continued during the allotted vacation, under the following agreement :—

For a dozen consecutive nights, the merchant was to be supplied with whisky in quantities sufficient to produce intoxication, and his companion was to keep sober, in order to protect their property and themselves, and to do all the offices of cook and "maid of all work." The last day and night, were to be spent in "sobering up," and getting things in order for a return to the duties of the head of a family, and the head of a mercantile house. At no other time in the year did this gentleman indulge in strong drink, and it was the only

specified time when the fisherman was under bonds to
keep sober.   Before leaving home, an estimate was made
of the exact quantity of spirits needed for the whole
time, and no more procured.   So that the fisherman was
obliged to be exact in portioning his rations, and a check
was thus placed upon himself.

Dr. Alexander Peddie, of Edinburg, who has contributed valuable treasures to the literature of inebriety,
says: "At the prison of Perth, one woman was committed one hundred and thirty-seven (137) times, for
being drunk, and when drunk, *her invariable practice
was to smash windows.*

"A man, when drunk, stole nothing but Bibles, for
which offence he was transported.   He was a soldier,
and had been wounded in the head.

"Another stole nothing but spades; another, nothing
but shawls; another, nothing but shoes; and a man was
transported for stealing tubs, seven different times; when-
ever he stole, he was, with one known exception, always·
guilty of tub-stealing."

That stealing should be associated with intemperate
drinking is not remarkable nor unusual, but the uniformity of the act under the influence of strong drink, is
exemplified by a sufficient number of cases to give
coherence to the theory, that there is a disease of the
brain in such persons, on which but little light is thrown
by our present knowledge of cerebral pathology.   We
are not informed whether these persons were conscious
of their stealing at the time of the act, or whether the
act was recalled by the memory, on a return to sobriety.

Like many other unsolved problems that are continually presenting themselves in the etiological study of such cases, their complete and satisfactory solution must be left to the methods of discovery and analysis that await us in the near future.

Dr. Crothers, of Hartford, and the late Dr. Beard, of New York, have recently made interesting inquiries into the Trance State in inebriety, and in reviewing my notes of cases treated a decade and more since, I find examples that are illustrative of the same variety of cerebral automatism, and I select the following in confirmation of their theory :—

CASE No. 29. A sprightly, attractive, good natured youth, fond of society and of sport, and especially of fast driving and trading in horses, of which his father and brothers owned quite a number. Residing temporarily, fourteen miles from a populous city, and visiting the city on parole for a day, he found it very easy to forget his obligation by indulging in drink.

When his intoxication had reached a certain point, which can neither be defined nor understood, the usual impulse to seek the stable, order a horse for a drive in the park, came upon him, and being a considerable distance from his father's stable in the city, he deliberately unfastened a horse that was secured to a post on the sidewalk, mounted the vehicle and started for the fourteen miles' drive to his temporary home. He avoided the carts, drays, carriages and cars that crowded the busy thoroughfares of the city, crossed bridges, passed turnpike gates, and in due time arrived without injury or mishap at his destination. Fastening the horse at the proper place, on arrival, and reporting himself in the house, he could give no account of his strange proceeding. He knew he was at home, recognized familiar faces, retired to his room, and on the following morning was as much surprised as any one, with the fact that he had driven home instead of taking the train, for which he had a return ticket in his pocket. An advertisement for the missing horse was found in the morning paper, and the young man, with an attendant, promptly returned the animal to an anxious physician, who was visiting a patient at the time of the disappearance. Regrets, explanations and apologies were pronounced in due order, of course, with an offer to make restitution for the delay and inconvenience, but the doctor, appreciating the situation, and grateful that his favorite mare was not damaged, dismissed the youth with wholesome advice.

Dr. Crothers mentions a case somewhat analogous, the man being "repeatedly convicted of horse stealing," and who was punished by imprisonment, and died of pulmonary consumption during his incarceration. His mother was insane, and his father " a weak-minded man." In my case I could trace no insanity or phthisis in the immediate family. Such cases are full of interest in a medico-legal, as well as in a strictly scientific sense. The time is at hand, when they will come oftener to the surface, and furnish material for study and discovery. These transcoidal states are, however, not new. They have not been observed as they might have been, for the

reason that they have simply been grouped with other
and more important symptoms, which so frequently
occur in fevers,—in hysteria and other of the neuroses, and
in which we are accustomed to regard such automatic
phenomena as indicative of sympathetic or functional,
rather than of structural disorder. Now that they are
being dislocated from their groupings, and studied
specifically, they stand out with a prominence that is
significant of more certain cerebral disturbance. What
the natural position of these phenomena is, may be inex-
plicable, in the same way that all the phenomena of life
are mysterious, and all attempts to trace them to their ulti-
mate cause, have been unsatisfying. And yet admitting
this as true, we are privileged to be at large in the vast
field of research, from which may yet be gleaned more positive truth. Enough, certainly, is apparent to the most
superficial and unscientific observer, to warrant the belief
that all such cases as have been described under the
head of disease, are beyond the reach of what are called
moral means. And certainly they are far removed from
the domain of law. They must be relegated to the realm of
science; and reformers, and advocates of temperance especially, can never reach the true level of the subject they
have in hand, till pledges, resolves and legal enactments, as
cures or preventives of inebriety, are abandoned, and
the intelligence of those who move in this direction is
cultured by a knowledge of the true causation, and career
of drunkenness. Some of the most prominent and
frequent immediate causes of intemperance, are never
referred to in the popular literature of the subject, and

certain complications of other morbid states which contribute largely to hinder recovery, are always overlooked
by authors and speakers; and until the researches of a few
pioneer thinkers have made them prominent, the facts
connected with the trance or automatic condition, of
which alcohol is the chief factor, would still linger
among the things doubtful and obscure. The following
remarks by Dr. Crothers are so fresh and original, as
bearing upon this phase of the subject, that I cannot
forbear quoting them :—

"The gospel temperance meetings, where the excitement is intense, are excellent places to study this
trance state. Men in different stages of alcoholic excess,
will come forward and sign the pledge, and manifest
great earnestness, and yet next morning be utterly
oblivious to everything done. I have seen many inebriates just this side of stupor and muscular paralysis,
be attracted to these gospel meetings by some means,
and become the most enthusiastic men, sign the pledge, ·
and describe their past degradation with evident satis-
faction, and close with the wildest assertions and promises for the future." * * * "Thoughtful men
often wondered why the same persons, who were so
enthusiastic, did not appear more than once or twice at
these gatherings. On inquiry it was found they had
fallen, or gone back, when in reality this was a trance
state, from which, on recovering, they did not return,
because they could not realize the position which they
· had taken. A United States Senator, who was an inebriate, appeared at one of these meetings, and made a

solemn pledge not to drink again; the next day he
denied it, and never believed or acknowledged that it
was true."

Doubtless there are some men, who, on discovering
what they have done, make an honest effort to keep
their pledge and lead a new life. Others, finding themselves in a false position, for very shame, avoid all efforts
of the kind afterward.

"In one case, a man who had signed the pledge was
for many weeks both speaking and drinking, creating
intense excitement everywhere, then recovered, with no
connected memory of it." * * * "It may be stated
as a fact, that in all cases where religious emotion is
constantly appealed to as an element of cure, the patient
is on the border of the trance state, and, furthermore,
any system of treatment which depends exclusively on
the religious element, cannot build up healthy tissue or
restore defective brain force."*

I think we are indebted to Dr. Crothers for being
the first writer on this subject, to give such prominence
to the trance view of it, and though his views, as just
quoted, may not be accepted as sound by many, they
are nevertheless worthy of careful thought, by all who
desire the whole truth.

Among the exciting causes which are not even mentioned in the popular literature of the subject, are
injuries, chiefly of the head; and I shall now present
some important facts, taken from a "Statistical Report

*See "Trance State in Inebriety, in its Medico-legal Rela-
tions," by T. D. Crothers, M.D., 1882.

of two hundred and fifty-two cases of inebriety, etc.,"
by my late esteemed friend, Dr. Lewis D. Mason, of
Brooklyn, and for the purpose of giving such cases a
definite place in the classification, I shall designate the
variety referred to, as

TRAUMATIC INEBRIETY.

Dr. Mason had ample opportunity for the study of
this subject in the wards of the " Inebriates' Home," at
Fort Hamilton, Long Island, and to his carefulness of
observation, added to the cautiousness of his nature, we
are indebted for the lucid and forcible classification he
has given in his " Statistical Report," etc.   Under the
heading, " Exciting Causes," the Doctor places first on
the list, " Head Injuries."   " At least one in seven of
the 252 cases (36) *became inebriates* from blows on the
head."   Mark the expression, " became inebriates."
They were not so before the injuries were received, but   .
because of the injuries, though most of them had previously used alcoholics in moderation ; 12 were fractures
of the skull; in four there was loss of bone; 22 had
concussion; 27 of the 36 'became habitual, and the
remainder periodical dipsomaniacs.   The Doctor remarks
further, that other diseases or injuries, such as are liable
to produce disorder of the cerebro-spinal axis, may either act
as exciting causes of inebriety, or tend to protract it. A case
is mentioned in which periodical dipsomania was associated with stricture, which, on being divided and cured,
resulted in removing entirely the attacks of periodical

drinking. Two other cases are mentioned, one of tape-worm, and one of necrosis of the bone of the leg, which being cured, "the attacks of dipsomania were not re-peated." Such examples of reflex irritation are quite commonly observed, as connected with other diseases, and as long as inebriety was excluded from the domain of pathology, and not considered as a disease, such remote influences were not supposed to affect the cerebral func-tions in connection with inebriation. Among the com-plications usually attending inebriety, there is no more common disorder than pulmonary consumption. In the family history, insanity, epilepsy and phthisis form a triple chain, the existence of which may be traced through several generations.

It is remarkable also, that among the complications attending inebriety, there is a variety of surgical troubles, as fractures, concussions, shot wounds, hernias, etc., which contribute freely of material to the tragic history of many such persons, and which are so conspicuous in retarding recovery, in many cases.

The following additional testimony in support of the doctrine of disease, is offered from the Blue Books of the House of Commons, which, not being accessible to most readers, is all the more valuable :—

Dr. Robert Boyd, F.R.C.P., of London, and Physician to the Marylebone Infirmary and Somerset County Lunatic Asylum, says :—

"Intemperance is the *outcome* of cerebral disease."

* * * "Intemperate habits may have been *induced* by cerebral disease."

"Of sixty-eight cases examined, which died in the Somerset Lunatic Asylum, the majority had disease of the brain, *prior to their continuous habits of drink.*"

"Persons of an irritable frame, and a nervous excitability, are generally more prone to fly to stimulants than others." * * * "In a great many cases symptoms of insanity precede the drinking, but I should think, in the majority of cases, it is the reverse."

Mr. William White, Surgeon and Coroner for Dublin, says :—

"When a man is apparently relieved from the direct effects of drinking, there remains the craving, which he cannot account for, and which he cannot control."

"Habitual drunkards have the desire themselves to refrain from it, and they make ineffectual efforts to abandon the drink; but the craving comes on, which they cannot resist."

Dr. J. Critchton Browne, the Lord Chancellor's Visitor of Lunatics for England, says :—

"Dipsomania consists of an irresistible craving for alcoholic stimulants, occurring very frequently periodically, with a constant liability to periodical exacerbations, when the craving becomes altogether uncontrollable. I have known it produced by injuries to the head, in perfectly sober and sedate men."

"If the drunkard is predisposed to nervous disease, or has any tendency to insanity, he inevitably becomes a dipsomaniac. If he has a strong nervous constitution, he may resist the poison for many years, without his mind being permanently affected."

Dr. Alexander Peddie, F.R.C.P., of Edinburg, says:—

"They are governed by an imperious impulse which they are *utterly unable to resist.*"

"He (the inebriate) is a slave of a passion from which he cannot free himself, struggle however hard he may."

"I consider it greatly in the nature of an internal disease."

"I know it to be a fact that many *children* who have been brought up without the use of alcohol, either in the way of nutriment or medicine, have shown the tendency to drink before they had any opportunity of learning the habit."

Dr. Francis Edmund Austie, Lecturer on Medicine at the Westminster Hospital, London, says:—

"Dipsomania is a disease which comes upon men at intervals, who are otherwise not inclined to drink at all."

"Periodical or paroxysmal drinkers have a great horror of drunkenness, in their sober moments. Compared with any other class of drunkards, they are far less blunted in their moral sense."

Dr. Arthur Mitchell, Commissioner in Lunacy for Scotland, says:—

"Dipsomania occupies a peculiar relation to drinking, being sometimes the cause and sometimes the product of it."

Mr. Balfour Browne, Barrister at Law, in London, and author of a work on "Medical Jurisprudence," says:—

"I believe that the dipsomaniac may not have the power of refraining from stimulants."

Dr. Forbes Winslow, of London, author of "The Brain and the Mind," etc., says:—

"In the majority of cases of habitual drunkenness, there is associated with it a disordered state of the brain which you do not cure. There is a disordered appetite which you do not eradicate. Although you keep the patient from drink, the craving for it is sure to return. There is no class of affections which, viewing them as mental affections, are so liable to relapse, as drunkenness."

Dr. David Skae, Physician to the Royal Edinburg Asylum, says:—

"Dipsomaniacs lose all control over themselves, and drink to any extent possible. In fact, they will drink anything they will get hold of, and if they cannot get spirits, they will drink hair-wash, or anything of the kind. I have known a lady to drink shoe-blacking and turpentine. These cases are mostly hereditary. They are very often caused by disease, by blows on the head, sometimes by hemorrhage, sometimes by disease of the brain. All these cases I mention to show that this is really a disease, and not mere cases of drunkenness."

Dr. Robert Dewitt, of London, says:—

"The habit of secret drinking is not amenable to moral and religious influences; it defies them. I have known many instances of women, amiable, respectable, and preëminently religious, who nevertheless were the victims of this habit, from physical or moral causes."

"Many of these persons suffer originally from disease, from a peculiar state of the nervous system, or from some insatiable thirst for these drinks."

Since the above has been in press, the author has received a very interesting and valuable letter from the Hon. Wm. S. Peirce, of Philadelphia, from which, by his permission, the following extracts are taken:—

The Judge says:—

"Some years ago a gentleman with whom I was very intimate told me some of the secrets of his life. I knew that he drank, but at the time he did not appear to be in the slightest degree under the influence of liquor. A few days after I referred to the matter, and he appeared to be wholly unconscious that he had ever told me, and said, with evident surprise, 'Did I tell you that'?"

"Another case was that of a member of the Philadelphia Bar. He was an intelligent man, but much given to drink. He had some important business to attend to before a judicial officer, and fearing he would not be able to attend to it, he requested a legal friend to attend to it for him. His friend went at the appointed time, and found the lawyer insisting on taking charge of the case, but evidently incapable of doing so. His friend managed to take control. A few days after, the lawyer called on him to know the result of the case, and what had been done. He had no remembrance of having

6

been before the officer, and had no knowledge of what had occurred there."

"Another case was that of a man who was convicted before me of stealing a horse and carriage. He had been drinking, but was not so much under the influence of liquor as not to appear to act intelligently. On the witness stand he testified he knew nothing of what had occurred from the time he began drinking till he found himself in the cell at the station house the next morning."

"Another case was that of a young man, about eighteen years of age, who was convicted of murder in the second degree before me. He was unused to drink, but went to a barn-raising and drank some cider. After that he went to a saloon and took a single drink of liquor, gin, I think. He had a pistol, and went roaming through the streets, apparently without purpose. He saw another young man standing against a fence, and without quarrel or provocation, deliberately shot and killed him. When he recovered from his condition, he was wholly unconscious of what had occurred, and did not seem to be drunk when these events took place."

These are typical cases, illustrative of the same condition as some of those quoted by others, and are important, as confirming, from a judicial standpoint, what has been noticed by physicians; and it seems to me that such cases, coming frequently before the legal mind, must eventuate in a modified jurisprudence, that will eventually secure a larger share of justice to a deluded portion of the race.

Judge Peirce very wisely says, in relation to them :—
" I am very much inclined to think there was some-
think like insanity in these cases, differing from the
ordinary and usual effects of drunkenness."

HEREDITARY INEBRIATES.

There is no research within the scope of biology that is
more attractive and instructive in its bearing upon inebriety than what is called *heredity*. I shall offer testimony to show how direct is the apparent transmission
from parent to child ; and yet the term heredity should
not be restricted in its meaning to this single form of
transmission only, for it is equally true that intemperance crops out in some families, where there has been no
previous lodgment of the taint, but where other disorders have been active, from which inebriety is simply a
deviation from the direct line, and is an evidence of the
beginning of the alcoholic diathesis. That under the
operation of the same law it has its termination also will
be shown, for the law of heredity recognizes periods of
limitation, as a necessity for the continuance of the race.
If it was not for such a law, and the degenerative processes
were to be continuous, without deviation or exhaustion, the
reproductive powers would sooner or later terminate,
and the race become extinct. It must be borne in mind,
therefore, that the union or mingling of certain temperaments, or qualities, may be so incongruous and inharmonious, as to create for the first time in any family record,

the starting point of disease. The links of insanity, epilepsy and inebriety, frequently alternate in the same historic chain, while pulmonary consumption, perhaps as frequently as any other disease, seems to lay the foundation for other morbid phenomena, that are manifested in the form of neurotic disorders. The likeness of progenitors may be developed sparingly or otherwise, and yet their completeness depend largely upon environment of the individual, and largely, too, upon training and culture. As a defective organ may, by use or disuse, be improved or injured, so may any impulse, appetite or craving, when it becomes manifest, be measurably subdued or stimulated by exercise and training. The discipline of organ or function by environment, is of itself, a study that should command the earnest attention, especially of all those whose structure is not in harmony with external surroundings. Not only is there transmission, but transmutation of disease, by heredity. Inebriety may descend as inebriety, but it is just as likely to change the form of its appearance into insanity, or other allied morbid manifestation.

Chorea may descend as chorea, but it is just as likely to appear in the progeny as epilepsy, the chief difference in the two forms of disease, being in the fact that the convulsive movements in epilepsy are paroxysmal, and in chorea they are not, while both may be identical in pathology. As water is changed into ice, so one disease may be transmuted into another form, and be known by a distinct name, at the same time possessing the same original qualities.

Amid the constantly varying and complicated dangers that beset the path of human life, and which constitute the environment of human beings, we should not expect completeness in the equally complex co-relations of the human body, which is so influenced by such surroundings. Taking these extrinsic factors of disorder into account, with the defective vital processes of nutrition, and other functional operations, and placing them in conjunction with the toxic vestiges of previous maladies, and it is not difficult to apprehend how inherent morbid predispositions may be aroused, and made to declare themselves with a potency that establishes their claim to inherited taint.

Dr. Wm. C. Wey, of Elmira, New York, thus forcibly presents the case :—

" It is well known, that in the searching review of the history of certain families from a fixed starting point, through lineal and collateral branches, no evidence of inebriety can be found.

" Men and women who represent this family, or more properly speaking, this idiosyncrasy, will, under all conditions and circumstances of bodily pain, mental suffering, pecuniary loss, affliction and disaster, preserve themselves from the consequences of inebriety. It is not through power of will, resolution, superior wisdom, foresight, caution or merit, that this exemption is manifested. Such persons could not become drunkards if they were so disposed. A certain moral and physical predestination, if I may thus use the word, protects them from the hazards of inebriety. No particular grace or excellence attaches to them for avoiding or escaping the consequences of alcoholic indulgence, although grace and excellence may often be assumed as the instrumentality by which they are spared the evils into which others appear so easily, and without let or hindrance, to fall."*

The fact of exemption from inebriety in a family, however, does not exempt from other conditions which may result in inebriety. There may be insanity, chorea, epilepsy, hypochondria or other taints, which in the order of descent may take the form of drink craving, or even crime. In the stream that flows from one generation to another there are collateral feeders that are continually pouring in their morbid products, thus modifying the old and eliminating new morbific forces, which take on a variety of forms.

From the report of Dr. Mason, already referred to, the family history is given of 161 of the 252 cases that he has so carefully tabulated, with the following result: Ninety-four had inebriate fathers, two mothers were inebriates, and both parents of four, grandparents and other relatives, 16. Insanity was traced to fifteen families, preceding inebriety. In some cases insanity and inebriety were present in the same family, showing the close and interchangeable relation of the two diseases to each other. There were also a number of cases of atavism.

From the records of three hundred and sixty patients

* Vide "Transactions of American Association for the Cure of Inebriates," 1871.

admitted to the New York State Inebriate Asylum up
to a certain date, the late Dr. Dodge, then superintendent, reported "forty-two as the offspring of intemperate parents, or one in eight. Thirty-six had intemperate
fathers, or one in ten. Six had intemperate mothers,
or one in sixty. Nine had intemperate brothers and
sisters, or one in forty. Sixty-six had intemperate
ancestors, exclusive of parents, on paternal side thirty-
six, or one in ten, on maternal side, thirty, or one in
twelve."

These are instructive figures, and yet in considering
the heredity of inebriety, as in other diseases, it is proper to go beyond the direct line of special disease, and
estimate the combination of temperaments which produce an alcoholic diathesis, without any previous alcoholic history.

As has already been said, it is by such union of qualities that the alcoholic craving is often established for
the first time, and the master proclivity begins to assert
itself, by what may be called a factitious heredity, that
engrafts the new passion as an anomaly upon the family
stock.

But let us look for a moment to the other fact,—the
great conservative fact of the law of hereditation,—that
such propensity may not only begin anew, in any given
family, but that it has its terminal point, as well. There
is a natural tendency in our very life, to find a settled
channel in which its forces may flow. So conscious are
we, in the earlier periods of our adult life, of the existence of strong and opposing passions and tendencies,

that we are apt to accept as an axiom, the common
expression that "character is not formed till forty."
Physical factors are all the time at work to develop
and mature what is in, and of us, and it is not until
maturity is reached, that a definite and uniform channel
is wrought out and established. This age of maturity
is not, and cannot be determined, according to a uniform
rule, but just as the sexual instinct has its time of beginning and ending, so other elements have a similar
history.

There comes a time in the course of one's life when
the forces that have been engaged in structural repair
and waste come, as it were, to a standstill, and consult
together as to which shall be dominant in the future.
Some organs of the body may be said to rest entirely,
after a certain age ; rest by a cessation of function, and
become atrophied also ; some, that do not rest absolutely,
undergo a modification of functional activity. It is at
this slack-water period,—at the middle of life,—when
the molecular deposits in the organic structure, are
different in quality than formerly, that we look for a
different product.

The craving for drink is not a craving of childhood.
It does not usually declare itself till the demands on
the nervous system begin to be exorbitant, and its
terminal period comes with as much certainty as does
its initial stage. That terminal period is the climacteric
period. A dozen years ago, or more, I called attention
to this thought, and urged those who had opportunity to
observe, to note the period of life when the largest

number of reformations, or cures of inebriety, were accomplished, and stated then that I believed they would be found to be between the ages of forty and fifty years.* Subsequent observation has confirmed this view. Between these ages, especially, do recoveries that are spontaneous occur. Statistics show that by far the greater number of persons first exhibit the alcoholic proclivity, between the ages of 15 and 25; and though I am unable to verify the statement by figures, I am convinced, that the allowance of twenty-five years of use will, in most cases, close the drinking career, either by the exhaustion of the desire, or by the fatal termination of the individual life. I think it will be found, also, that when inebriates have lived beyond the middle period, so as to reach the three or four-score limit, the commencement of the drinking career was considerably later in life than the average period named. There is a good reason for increased longevity in cases of this sort. The later the propensity to drink develops itself, the nearer the person is, to the period when the constructive quality of the metamorphic processes, that are essential to living, gives place to the destructive, in which cases the use of alcohol is not only admissible, but desirable; for if there is ever a period in human life, when alcoholic beverages are indicated, it is the period of senile degeneracy.

I have referred to environment as an important force in the discipline of the functions of the human body, and now call attention to the importance of a practical adjustment of individuals with an inherited tendency to

* Vide Proceedings, etc.

alcoholic indulgence, to the conditions and circumstances, amid which they find themselves placed.

So far as regards the perpetuation of the human machine, and of all the organs and functions which are essential to its repair, nature has provided within the body itself, a sufficiency to reach the end intended—the ultimatum; but it must be remembered that the ultimatum is not uniform; it differs as widely as human constitutions; but, as Dr. Dobell puts it, " there is also included within the functions of the organism a power of *disposing of poisons,* so as to avert their fatal effects, but this power is restricted by certain limits." *   *
" There is, then, no room to doubt that the same force which protects the organism against the undue influence of poisons generated within itself, protects it also against the poisons introduced from without."* *   *   *
" Some of the force manifested in development, growth, maintenance and repair passes on into the germs of new cells and tissues, and again is passed on from the *entire* . *organism* into the germs of a new generation." *   *
" It is reasonable, therefore, to conclude, and to affirm, if the existence of this force is recognized, with its power of succession and transference, that the constitution of the being possessing it, both in material and force, will be dependent on the conditions of the world in which it exists, subject to alterations, co-relative with that external world itself." Thus it is evident that the individual who is conscious of an inherited tendency to alcoholic

* " Lectures on the Germs and Vestiges of Disease;" by Horace Dobell, M. D.

excess, may do much to modify, if not to control its force, by placing himself under such conditions of living as will tend to increase his constitutional vigor in the direction in which it is most needed. A person coming into the world with a tendency to pulmonary consumption will, as soon as he knows it, begin to co-relate himself with the most favorable conditions of climate, occupation, etc., that the progress of the morbific element within him, may be arrested, if possible, and that the normal forces that are antagonistic to this manifestation of disease, may be strengthened.

The same is true of other disorders. It is conspicuously true with the disease of inebriety. The fact of an inherited tendency to it, and the other fact, that by reason of such an inheritance, he is fortified with a potent stimulus to resistance, is an influential factor in the direction of health. It is lamentably true, however, that the popular view of this fact of inherited tendency, has influence in the opposite direction. The opponents of the doctrine of disease, insist that the knowledge of the fact of inheritance, is sufficient to arrest all effort at recovery, and is equivalent to abandoning the victim to a fatal doom. There is nothing more illogical or irrational in the entire field of polemics, as related to this subject, than the assertion so commonly made by reformers, that the dogma of inheritance leads to fatalism and despair. It is the great conservative truth in relation to other physical conditions, which all men may avail themselves of, who are inheritors of any special predisposing morbid element. If they know of its existence, they are fortified, by that very fact, with a weapon of resistance. They are on their guard. They fear to be overtaken without watchfulness,—and they watch. Knowledge of danger is the guardian power, on which they rely as the inspiration to effort, to harmonize surroundings with internal conditions, and thus avoid the risk consequent upon ignorance, or doubt. It is impossible even to approximate the number of inebriates by heredity, and those whose inebriety is due to external causes ; but I am satisfied that among the chief hindrances to recovery from a life of inebriation to a life of sobriety, is the false teaching of those, who overlook the aspect of disease, and limit their labors and appeals to the domain of morals and ethics. When society comes to learn that the cause of inebriety is primarily in the disturbed relations between different organs, and functions of the human system, and especially, that children come into the world, bearing with them the vestiges of disorders that have lingered through one or more previous genera- . tions, light will begin to reflect its brightness upon new and improved practical methods. When society comes to add to this fact of inheritance, the other fact, that by adapting the educational, social and hygienic surroundings of the growing generation, to its physical necessities, much may be done to arrest or counteract the development of the craving for drink, additional lustre will be imparted to the recuperative and restorative measures, that will appear as new revelations to the advanced intelligence of such a period.

The following quotations from witnesses before the Habitual Drunkards' Committee, of the British House of Commons, confirm the doctrine of heredity.

Dr. Francis Edmund Anstie, says—

"The tendency to drink is a disease of the brain, which is inherited."

"When drinking has been strong in the families of both parents, I think that it is a physical certainty that it will be traced in the children."

"I believe if a man habitually took two or three times as much stimulant as is required for the physiological wants of his body, his children would inherit those affections."

"I desire to be understood that my opinion, based upon the most recent physiological researches, is, that alcohol in moderate quantities is an exceedingly useful article of daily food, at any rate for large classes of the community."

Dr. J. Critchton Browne, of London, says—

"I think that the hereditary tendency from drunken parents does not always manifest itself in insanity, but frequently in idiocy and crime."

"I think that in about one-third of those cases of insanity which I found to be due to drunkenness, there was a hereditary tendency, as a coöperating cause with the drunkenness, creating the predisposition."

Dr. Forbes Winslow, the distinguished teacher and author, says—

In examining a list of criminals "there was a father a drunkard, a grandfather a drunkard, grandmother an idiot, and in the whole line of that family there were drunkards, there were criminals, there were idiots; all forms of vice were hereditarily transmitted."

Dr. Alexander Peddie, of Edinburgh, says—

"I believe the habitual drunkard inherits the proclivity from drunken parents, or an insanity from a constitutional insanity in his family, in which the most marked manifestation is a tendency to drink."

"A lady of good education and good principles began to drink at the age of sixteen, died at the age of fifty-six, during which time she had many severe and protracted fits of drinking, and at last drank herself to death. * * When under control, she was intelligent, active and industrious, making herself useful to others. Sometimes she employed herself as Bible-reader, and when in the country three years, under continued control, she gave herself to geology and botany, and wrote most excellent letters from her retreat. Her father was an habitual drunkard, a grand aunt on her father's side . was the same, and also a cousin on the father's side. On the mother's side, the mother was an habitual drunkard, a brother was also, another brother was insane, an aunt was an habitual drunkard, two nieces, daughters of the same were habitual drunkards, and some other members of these two families are said to have been affected in the same way."

Dr. Robert Druitt, of London, author of a work on Surgery, says—

"The condition which gives rise to intemperance is hereditary."

"In some families there is a nervous system of bad construction ; the members of those families are marked by enormous consumption of both food and drink, and at the same time by an enormous oxydation and waste. They are constantly in a state of chronic hunger and exhaustion."

"The class that I appeal for are chiefly women of the upper classes, or men who are led to secret drinking for the relief of misery, bodily or mental."

"They have frequently inherited a very feeble or a very excitable nervous system." •

* * "When persons have imperfect brains, when they are eccentric and slightly subject to disease, that state is hereditary, and it will lead to drink."

Dr. Charles Elam, M.R.C.P., in his remarkable volume, "A Physician's Problems," presents some striking examples of transmitted tendency to intemperance, from which the following are taken :—

"The habit of the parent, when inherited, does not appear in the child *merely as a habit*, but in most cases as an irresistible impulse, a disease."

"This disease, known as dipsomania, is quite readily to be distinguished from ordinary intemperate habits ; it is characterized by a recent writer as 'an impulsive desire for stimulant drinks, uncontrollable by any motives that can be addressed to the reason or conscience, in which self-interest, self-esteem, friendship, love, religion, are appealed to in vain ; in which the passion for drink is the master passion, and subdues to itself every other desire and faculty of the soul.' "

" The tendency to suicide is frequently, though by no means invariably, allied to the heritage of drunkenness." * * " Four brothers inherited the passion for drink, which they all indulged to excess. The eldest drowned himself, the second hung himself, the third cut his throat with a razor, and the fourth threw himself *out of an upper window,*" etc.

Dr. John Nugent, Inspector General of Lunatics for Ireland, says: " I have a case of a gentleman of fortune ; the father was insane, he had a brother ; one of the brothers took to drink and the other became insane without ever indulging in drink at all ; so that the *hereditary* disposition showed itself in one by actual insanity, and in the other by drinking."

INEBRIETY AND INSANITY---HOW RELATED.

Nobody has yet solved the problem of insanity. No lexicographer has given it a definition that can be accepted as a rule of action. No scientist has expressed its meaning in language that the law recognizes as authority. No jurist has defined it in terms that science can comprehend. The reason is, that the eyes and understanding of law, are different from the eyes and understanding of science. And yet insanity, in its main features, is one and the same thing, and has been during all time. Its pathology is unchanged, and its symptoms are alike, and have been, from age to age. The law cannot see it at all unless it does something. It must rave in open madness, or skulk beneath the cloud of speechless melancholy. It must hurt or kill, it must steal or burn, or do some overt act, before the eyes of the law can even discover its existence. So, likewise, science, while it penetrates beneath the external and visible, and finds it lurking in mute disguise within the cunning network of the human frame, fails also to define it.

Inebriety, on the other hand, has been visible and offensive, from generation to generation. It has obtruded itself upon society as a disturber of domestic and public peace, and all people have looked upon it, from all sides and with all kinds of eyes and understandings, and it has been described with a comparatively uniform exactness, though its true definition and nature have not until recently been known. The law has not penetrated the hidden places of our being, to seek out occult forces that impel an inharmonious nature to indulge in alcoholic excitants; and hence drunkenness, like insanity, must do something before the law can apprehend it. It must steal or burn, commit an assault, or in some way disturb the peace and safety of society, or the law cannot lay hands upon it. Science alone has been able to discern the latent pathological state from which the drink craving comes, to command and control the being who is its victim.

Insanity is a disease of the brain, some say, of the mind. Viewing the latter as a distinct entity, but

7

avoiding in this place the discussion of the somatic and
physiological theories of insanity, it will be assumed, as
we know nothing of mind as an entity separate from
the brain, that insanity is essentially and exclusively a
brain disease.

It is not so with inebriety. Insanity means mental
unsoundness, in a general sense. In a more particular
sense, deprivation of reason. Hence no one is insane,
whose processes of intellection and ideation are regular
and complete, and whose organs upon which such processes depend, are in good working order. Inebriety
does not mean or imply mental unsoundness. Inebriates
reason as logically and clearly as persons of the same
degrees of culture, and with like opportunities, who are
not inebriates. Their processes of intellection and ideation, when sober, are undisturbed. When poisoned by
alcohol, they are like persons under the toxic influence
of any specific disease which affects the brain, and for a
time are off their poise, and become delirious, as in
typhoid fever, or when the blood is poisoned, or in
cerebral inflammations when the sensorium itself is
assailed.

Insane and inebriate persons may all be out of harmony with their external relations, but from different
causes, and under different conditions, and it is important
that the distinction should be drawn ; the mental vision,
the brain-perception of the insane, causes him to be
athwart his circumstances and surroundings. He does
not see things as they are. He does not comprehend
himself, and he maintains his position with a persistency

that characterizes his disorder.   The inebriate is out of harmony with his environments, only when he is inebriated, for it is important to remember that an inebriate may but seldom be under the toxic power of alcohol, and when he is not, his relation to others and to the world about him is a normal relation.   His disease is an irrepressible longing for the state of drunkenness ; not so much for the liquor that produces intoxication ; and he is just as much an inebriate,—diseased,—when he is able to control for the time his desire, as when he indulges.   The indulgence may be regarded as the second stage in the manifestations of his disease.   Until he yields his will, and gratifies his propensity, he is usually in an intellectually normal relation to his family and the community.   The surrender of his volition, and with it, his judgment and his interest, is the first evidence of the overpowering mastery of his passion, which prostrates and beclouds his reason, until his debauch is accomplished.   Then all his powers of mind re-assert themselves as before, and he is re-established in his domestic and civil relations.   I do not say that the inebriate is well balanced.   He is not.   He is apt to be fickle, uncertain, liable to sudden departures, and all that ; but these are not insanity.   They are not the product of impaired intellection or defective knowledge. He can say with the great master of poetry, Scott—

"I have been hurry'd on by a strong impulse,
Like to a bark that scuds before the storm,
Till driv'n upon some strange and distant coast,
Which pilot never dream'd of."

When sober, he regrets what he did that was wrong

while under the influence of liquor, and is anxious often to make restitution. The insane, on the contrary, do not excuse themselves for offensive conduct. They think themselves right, and are self-justified. Their reason does not apprehend the enormity of evil doing, and their judgment is incapable of determining ethical questions which are intuitively discovered by intelligent inebriates.

Inebriety being due to physical causes, the general belief is that what are termed "drunken fits" are induced voluntarily, and that the inebriate is therefore doubly to blame for any overt or irregular conduct; to blame first, for drinking to excess, and next, for whatever may be the consequence of his excess. It is indeed true that the will, by nature and right, is the supreme arbiter of moral conduct, but it is equally true, that it is sometimes overpowered and subdued by the drink-craving, just as we are overcome and subdued by extreme pain, or exhaustion, or shock. We suffer the pangs of disease, or serious accident, and it is but seldom that even the most stalwart frame, or the grandest intellect, can, by the mere behest of the will, overcome or modify physical suffering, or hold back the groans and tears, which are the normal expression of pain.

A man is in extreme danger,—shipwrecked at sea. He is clinging to a spar in mid-ocean, exhausted by anxiety, hunger and thirst, with no means of relief in sight, nor scarcely any to be hoped for. He has no pain, but he thirsts for water. He knows if he drinks the salt—sea—water that his thirst will be intensified, and

he will be made more miserable. But the demand
of his nature is for liquids. The fluids of his body are
diminished, almost exhausted, his tongue is parched, his
throat dry, and he must have water. He yields to the
craving, and as a dog laps water from the wayside pool,
so he, with his parched tongue, laps the waves as they
pass, thus increasing his agony and danger. He does it
because he must. His will is powerless. His condition
is analogous to that of the dipsomaniac. The demand
is physical, the suffering is not that of pain, but of an
unsatisfied yearning for relief. It may be a starved or
a diseased appetite. In either case the suffering is
intense. It must be relieved at any cost. In the same
sense, insanity is the result of voluntary action. Many
an unhappy victim of a morbid propensity, who scarcely
knows what it is, or from whence it comes, or how to
describe, and much less to define it, is afloat on the sea
of worldly pursuits, and human duties, clinging to his
manhood, and to the consciousness of his inalienable
title to a place in the family of men; and yet misunderstood by others, and failing to comprehend his exact
relation to them. He seeks the counsel of his physician, who, apprehending the situation, and foreseeing the
danger, interposes restrictions, and outlines a course of
living which he must observe, or fall into the pit which
his grandparents, or some other venerable representa-
tives of a past generation have dug for him. He is not
a physiologist nor a philosopher, but a busy, hard-
working, and perhaps money-making *genius.* He does
not discern the wisdom of the advice he has received ;

or, if he does, he does not see how the wants of his family are to be met by any change in his business, or his habits, and though the advice is emphatic, it is disregarded.

The result is insanity, in very many instances; but when the inebriate and the lunatic are both summoned to appear in the presence of the law, to answer, it may be, for the very same offence, the verdict in one case is likely to be voluntary drunkenness, and additional punishment for the offence, and in the other, insanity by the " visitation of God," and kindly care in an asylum.

It is a fact pronounced by the teachings of psychiatry, and well known to law and to medicine, that insane persons are frequently not only conscious of criminal acts, but of the penalties belonging to them, and that so far as knowledge may be considered a test of responsibility, they should not escape the decision or the consequences of the test. The principle of rewards and punishments, is recognized as a governing principle in the conduct of lunatic asylums. Patients are removed from one ward to another, restrictions are imposed, and indulgences granted, as means of discipline. The higher motives and faculties are appealed to, the weaker ones strengthened, and those which are excessively developed or abnormally displayed, restrained. This is a rational mode of treatment, which has commended itself to the judgment of our wisest alienists, from the days of Pinel and Connelly until now. Why not, therefore, deal with mental unsoundness outside of asylum walls, in the same way? If insanity deprives a person of self-consciousness and reduces him to oblivion, and perpetual darkness, he is, of course, no longer a responsible being, and but one alternative is left to him.   If, however, his insanity takes the form of an acute and temporary perversion of his moral sense, or mental faculties, but does not so far invade the sensorium as to deprive him of discriminating power, I see no reason why he should be exempt from a wise administration of penal discipline, the same as other violators of law, with perhaps less knowledge and discrimination, who, in these days of sentimentalism, and pseudo-philanthrophy, are so unfortunate as not to be adjudged insane.   The busy merchant, the over-taxed professional man, the speculator in politics, morals or stocks, too often rushes on to a painful doom, in the midst of the whirl and pressure, heedless of his own conditions and fears, and of the counsels of his physician.   Regardless of what they know to be the inevitable issue of such a course, such persons pursue it, until insanity, inebriety, or some other damaging disorder of the nerve centres, overcomes and enslaves them.   Where knowledge is possessed of the existence of any morbid predisposition, and of the danger of encouraging its development, its possessor fairly earns the reward which he will surely receive, if he allows his volition to guide him in the path that leads to his own destruction.   Who shall measure the degree of responsibility, in either case, by any other standard than that of knowledge, and the power of choice?

One impressive fact, in connection with the inter-

change of insanity and inebriety, should be empha-
sized in this connection. It is this: that a drunken
carouse not unfrequently arrests and forestalls an impending seizure of insanity. For example, take a family
in which various neuroses are exhibited. There are two
brothers, who hasten and worry with business cares and
responsibilities, until they are suddenly aroused to a
consciousness of danger. Headache, wakefulness, indigestion, irritability of temper, or some kindred symptoms, are manifested. They seek advice, and a change
of scene, and rest, are recommended. One of them
stops, and the other continues, as an apparent necessity,
to conduct the business. The stopper rests, and may
naturally drift among idle or luxurious people, and discover a susceptibility to the alcoholic poison, which is
intensified by indulgence, until a profound debauch
occurs. Recovering from it, he finds that the complete
arrest and revulsion of all his functions, and faculties,
occasioned by the profound narcotism (intoxication) has
temporarily subdued the insane diathesis with which he
was born, and that he is enabled to resume his labors
with courage and assiduity. The other, suffering from
similar symptoms of mental depression, restlessness, etc.,
struggles on, without rest or change. No impulse to
drink manifests itself, or, if it does, from motives of
propriety and conscience, he may be able to subdue it,
while, from continued pressure and resistance, the hope-
less victim drifts away into the gloomy domain of mental alienation. Take the cases of two young lawyers,
both of whom I knew well.

No. 201, was brilliant and enthusiastic. His father died, a victim of over-work and alcoholic intoxication. His mother was fond, indulgent and unwise. As he came to his majority, with a moderate competency at his command, he chose the life of a debauchee, from a mere love for it. A few years, however, satisfied him that it was a life of wretchedness, and constantly impending ruin. In view of this disclosure, which came to his own consciousness, in moments of serious reflection, he resolved to stop. His will, fortunately, was by nature almost resistless, and under its guidance he was sometimes enabled to stop for a season. Yet at other times the old passion would return, with renewed violence, and he found the safer course, was voluntarily to place himself under restraint. The resistance he was compelled to command, in order to subdue his inward passion, together with the humiliation of restraint, though self-sought, wore upon his nervous forces, frequently to the point of exhaustion.

In this state of extreme depression of body and mind, his intellect seemed to reel upon itself, and he became clamorous in his demand for his accustomed drink, as a means of temporary relief. A moderate quantity of whisky, judiciously administered, with milk, would have restored calmness, and promoted rest for awhile; but again and again the demand repeated itself, till he resolved, with a clear head and an honest heart, to do without it, regardless of consequences.

On one occasion, however, during his struggles, violent symptoms of acute maniacal exhaustion supervened,

and so intense was the desire for drink, and so unre-
mitting his demand, even for a single draught, that it was
evidently a physical necessity to relieve a state of intense cerebral exhaustion ; but being refused by his
keepers, he fell upon his knees, and with imploring gaze,
trembling from crown to heel, died, in a paroxysm of
frenzy, his friends piously congratulating themselves
that he had died sober, with reason dethroned, rather
than afford him temporary relief, and give a possible
chance for recovery, by the indulgence of a single glass.

The other case is No. 86. A lawyer also, not brilliant,
like 201, but of average qualities and capacity. He,
too, was determined to martyrize himself, if need be, in
the struggle to overcome an inherited tendency. The
other surrendered his reason, and died in an asylum,
while No. 86 died at his home, with a mind sunken into
partial dementia, by the struggle of years to keep and
remain sober. He had been a deceptive, cunning inebriate, who would do almost anything, theft, lying,
deceit, to procure liquor, and then drink to the extreme
of helplessness. Year after year, he continued his potations, till his capacity for enjoying them was exhausted ;
and it became a matter of effort, to take a sufficient
quantity to realize its accustomed effects. In this stage
of saturation, when every organ and tissue seemed charged
with the poison, it was heroic to make an ·effort to redeem himself, seeing that his moral forces were kept in
abeyance, and his apprehension of methods and results
was in subjection to his sensual impulse ; but still the
struggle was commenced.

Under a strong conviction of duty, aided by doting parents and an ever watchful sister, he succeeded in abandoning his cups, but the cost of the effort was a complete depression of vital force, a sudden change of morbid propensity, an absolute loss of mental poise, and a sober death; the death of dementia. The family taint was not distinctly marked in the preceding generation, though it was evident, in certain idiosyncrasies and oddities, that it must some time declare itself in a specific form. In this case, also, an inborn proclivity for alcoholics terminated in death from insanity.

This interchange and transposition of morbid inheritance, in whatever form it may appear, is a significant fact which seems to teach that there is almost as much reason to believe that intemperance sometimes prevents as it does cause insanity. In each of these cases the mind gave way with the withdrawal of alcohol. Had its use been continued in graduated and *medicinal* doses, under the direction of a wise physician, the end might have been delayed, and death welcomed by an unclouded intelligence.

While engaged in writing this chapter my attention is called to a case as follows: A young man in humble life, a huckster by occupation, calls upon me with a relative, and an officer of the law, for advice. He is the son of a father, whose wayward life is notorious, and whose abiding place is never settled, whether from mental unsoundness, moral depravity or a love of vagrancy, is not known. The wife maintains the home as best she can, and the son, during a period of industry, sobriety, and filial duty, contributes his portion towards it. The time comes, however, when, from constitutional tendency, from excess of worry and of work, or for want of occupation, he becomes wakeful and restless, with loss of appetite. Failing to seek advice at such times, he drifts into despondency, and has twice been an inmate of an insane asylum. On other occasions, he has escaped the sequel of his wakefulness, restlessness and anorexia, namely, melancholy, by indulging in drink, and has for a time been a great annoyance to his family and friends, but with the cessation of his debauch, there has always been a return of capacity for work.

Indeed, the views of professional alienists have undergone considerable change, of late years, on this subject. It is found that intemperance is not the all-potent factor of insanity that it was supposed to be, many years ago. The facts of insanity, like the facts of crime, are exhibiting new and important physical causes, for numerous departures from typical standards of integrity of brain, nerve and blood. Civilization, and the civilization of the American people especially, is an energetic force in the creation of new methods of social life, by the effective operation of which, new discoveries are being made of profound departures in the functional activities of our bodies, from the recognized normal standard.

This fact is frequently shown in an evident instability of the great nerve centres, which have to do with the processes of intellection, idealism and nutrition. Such disclosures are, of course, making their impression on civil law, which is constantly being modified in favor of the higher and more intelligent apprehension of hygienic and sanitary requirements. Witness sanitary legislation, by the creation of Boards of Health, with ample powers to do what the best scientific knowledge indicates, in the direction of public health.

Dr. David Hack Tuke, of London, than whom there is no higher authority on the subject of insanity, expresses himself thus, as to the relation of intemperance to insanity, and especially as to the alleged increase of insanity: " I do not deny that there are instances of persons whose mental condition is benefited by the use of a diet into which some form of alcohol enters, especially the light wines and beer. The misfortune is that the very people who are likely to be thus benefited are often to be found among those who, from noble motives, abstain from wine. Thus it comes to pass that the folly and wickedness of intemperance involves a double evil. Intemperance not only injures those who yield to this vice, but it leads many, by natural reaction and indignant recoil, arising out of the knowledge of such abuse, to deprive themselves of a beverage, or even a medicine, which might act in their case beneficially, in gently stimulating the functions of the brain, and lessening the tendency to nervous irritability and languor. Those who suffer from the want of a moderate stimulant, and fall into a depression which might have been warded off by their use, are, I maintain, the victims of intemperance, and their discomfort or actual insanity lies at the door of the drunkard."*

* " Insanity and its Prevention," by D. H. Tuke, p. 203.

If this is a fair statement, then it is evident that abstinence from alcoholic liquors by persons whose health would be benefited by their moderate use is, or at least may be, a factor in the production of insanity, and it also confirms the observation I have already made, and illustrated by the clinical record of the death of the two young lawyers.

The alleged increase of insanity and inebriety, as well as other forms of nerve and brain disorder, together with their relation to human conduct and domestic happiness, has become a favorite theme for study by thousands of intelligent persons, which is a most cheering sign of the times. The more thought, and the more light, that can be brought to bear on this subject, the more good will be accomplished. Science has never, until recently, found her true place, especially in the realm of psychology. She has quietly, and unobserved, been plodding her weary ways in the direction of other pursuits, while religionists and philanthropists have, seemed to possess the entire field, if not of psychiatry, of mental and moral unsoundness, as connected with the habit of dram-drinking. But this is not sufficient. Scientific thinkers are now bringing together inebriety and nerves; insanity and brain; crime and heredity; law and justice; all in their own order and relations to each other and to society at large.

Whatever may be said of popular prejudice and ignorance of the fixed routine of legal proceedings, and of the implacable repulsion of popular theology, to advanced views and methods, we are nevertheless making progress in this matter. Society is beginning to recognize the causation of certain human conditions, which have hitherto been misunderstood, but which are now emerging from chaos, and asserting their claim upon public attention, and which must lead to reforms in jurisprudence, that shall be consistent with its own highest aims.

In a strict ethical sense, there seems to be no difference between what is called voluntary inebriety, and what may justly be called voluntary insanity, or voluntary gluttony, seeing that excessive drinking, excessive working, and excessive eating, all spring from allied nerve conditions, and are alike productive of diseases. If we look backward and trace the lines of their descent, we shall generally find that each has its antecedent factor in some ancestral fault. A patient presents himself for advice, his constitutional tendencies, his family history, and his modes of life being all well known to the physician who is to prescribe. The physician informs his patient, that he is deliberately violating the laws of his being, so that the normal harmonies of his functional life are thrown out of gear, and that unless new methods are adopted, his latent predisposition will develop and assert itself, and that the inevitable result will be insanity, or some similar disorder, which may disqualify him for social, domestic or business life. He thinks, however, that he cannot afford to relax his routine of daily service. He cannot abandon long established habits of living, and he rushes on, contrary to advice, till his family weep over him at last, as a

lunatic, an inebriate, or it may be, a glutton, dying from
apoplexy or paralysis. It matters a great deal, however,
in public estimation, how he dies. It matters much to
the reputation of the family, whether the ban of intoxication is upon him, or whether he leaves this world
because of violating the laws of his being, in the way of
overfeeding or mental strain. It is, however, true that
insanity may be frequently prevented, even more readily
than inebriety. A person with a direct hereditary taint
of insanity, may pursue a course of life, under professional
guidance, which will secure him against a public exhibition of his infirmity, with more ease than a person
with an alcoholic diathesis, can be kept from indulgence
and exposure. In the former, there is no physical
craving to overcome, no struggle with an internal and
positive demand, which, strong and imperious in itself,
is rendered more so by the allurements of social, and the
attractive displays of public life. All that is required
in the one case is, in the very onset, to submit to intelligent guidance as to mental and physical hygiene, so as to
preserve a normal equipoise. In the other, while it is
necessary to accomplish the same end by similar methods,
there is the additional task of keeping in subjection a
strong physical impulse.

Hunger and thirst cannot be averted by an act of
will, even in the most exalted health of body and mind.
Working, walking, thinking, and all other acts that are
muscular or physical, are more under control than strong
physical cravings. The late distinguished Forbes
Winslow, of London, in writing of such minds, says :

" We read the sad, melancholy and lamentable results
of either total neglect of all efficient curative treatment,
*at a period when it might have arrested the onward
advance of the cerebral mischief* and retained reason
upon her seat, or of the use of injudicious and unjusti-
fiable measures of treatment under mistaken notions of
the pathology of the disease. * * * Experience
irresistibly leads to the conclusion that we have often in
our power the means of curing insanity, even after it
has been of some years' duration, if we obtain a thorough
appreciation of the physical and mental aspects of the
case, and persistently and continually apply remedial
measures for its removal."*

There is seemingly too much sentimentalism about
this dogma of insanity. It is environed in the public
thought, with a host of sympathies, which, while they
are alike human and humane, are, nevertheless, hindrances, often, to a proper recognition of its true nature. In
many a family, there are those who typify inherited
tendency, showing itself as neuralgia, hysteria, inebriety
or insanity, all of which are sometimes seen in a single
household. Take, however, a larger family, a community,
a town or village, and the same diversity may be observed
on a larger scale. They all have their places, and each
class its proportionate and appropriate place. They are
all deviations from a normal type of humanity, and all
represent a common origin ; they come from disordered
digestion, imperfect nutrition, impure blood, and enfeebled nerves.

* See Lettsomian Lectures, pp. 59, 61.

8

We have thus far outlined some of the resemblances and some of the differences, which are represented in insanity and inebriety, and while the distinctive marks may not be equally prominent to the view of all observers, a few of them are sufficiently clear to be recognized by all. In insanity, the judgment is not to be relied on. One can never feel safe in accepting the judgment of an insane person, even if the subject judged of, does not come within the limits of his hallucinatory sphere. Sometimes it is even stupefied—collapsed—abolished. The higher intellectual faculties may, however, be so slightly disordered, as to render it difficult to determine the imperceptible passage from a normal to a diseased condition. In either state, there is a sense of insecurity and danger. We would not for a moment, set out upon a journey in a vehicle with its wheels, or axles, or any part of its running-gear visibly out of order. We would also feel timid and apprehensive of accident, if we knew of any part of the vehicle, even if out of sight, was slightly damaged, either by accident or ordinary wear.

So, if we see in a fellow being, positive evidence of brain lesion, we cannot trust ourselves to the guidance of his judgment, even in trivial affairs. If, however, there are no visible signs of brain disorder, and we only know of the pressure of eccentricities and obliquities, which are sufficient to entitle their possessor to the *suspicion* of insanity, we are on our guard, apprehensive and expectant, to a degree that destroys confidence.

Again, the persistency of the insane mind in holding on to its delusions, is worthy of notice as a characteristic sign; persistency in the direction of delusions. It need not be always the same delusion, but yet in the same course. Delusions in inebriety, are transitory, and soon pass away. They are as frequently due to sympathies with remote organs, as to any immediate impression upon the cerebral substance.

Just at this point we may pause to recognize the difference between mere functional, and organic disturbances. There is nothing more common than to have delirium manifest itself in the career of certain acute disorders, as in rheumatism, typhoid and typho-malarial fevers, but the delirium in such cases, is symptomatic and sympathetic. It does not betray any lesion of the brain, or its coverings. It is an expression of transient nerve disturbance that may pass away in a few hours, unless its invasion is so complete and overpowering as to be itself a symptom of death. Delirium tremens is a disease that may be so classified, and as this disease is not confined to the consumers of alcoholic drinks, it is fair to assume that the theory of blood poisoning as the cause, does not hold good. It is more rational to assume that the toxic force of the drug is expended upon the nerve centres, and that the symptoms vary accordingly. Delirium in such cases is not insanity, as we generally believe and accept the term.

It has been shown in another chapter, that superintendents of insane asylums in their organic form, as an American Association, have distinctly proclaimed their position in this matter.

Their opinion is decided, that inebriates are not considered as insane, or at least not sufficiently so to be committed to asylums for the insane, where their presence is prejudicial to the latter class.

This declaration, coming from the source it does, is entitled to weight, and will have its influence.

Assuming it to be true, then, as stated by these expert alienists, that the presence of inebriates in asylums and hospitals for insane, is offensive and injurious to the legitimate occupants of such institutions, it simply narrows the foothold upon which the inebriate is obliged to stand, in making an effort to redeem himself from the curse that blots his name, and burdens his nature. Already repudiated by society, the church door closed against him, business interests kept from his grasp, and even the lowest ambitions crushed out of his struggling self-hood, with no hand to guide him when at large, and no hospital roof to cover him when debauched by excess, or raving from thirst madness, his relation to the world in which he lives, and to the local community of which he is one of the constituent parts, is certainly phenomenal, and demands explanation. Fortunately, there is one spot to which he has an inherent claim, and to which he may, in the large majority of cases, return, and that is, home. Unfortunately, however, his presence at home, is often dangerous to the home itself. When he goes to it inebriated, he carries with him fear, sorrow, shame, and sometimes, worse than all, penury and death. When not inebriated, home does not comprehend him, nor does it know how to protect him from himself and his secret foe, when sober.

The law is, of necessity, without heart or sympathy. It does not know what such qualities are. It would not be law, if it possessed them. It is inexorable. It knows no higher virtue, than obedience to its own commands. Its duty is to protect the injured family, or community, by the arrest and the confinement of offenders, but to offenders, it grants but little sympathy or moral help.

Temperance Societies succor drinking men, when they are sober, but in the hour of keenest strife with self, when the pledge that has been honestly taken, and the sense of honor involved in its violation, and every manly sentiment and principle, that dwells in human nature, all array themselves, as a shield to fortify and give conquering strength; in that moment, such conditions, and such memories, and such convictions, are as available to control a paroxysm of mania, to prevent a seizure of hysteria, to antidote a poison of malaria or smallpox, as they are of themselves, to subdue the mania for thirst in the constitutional inebriate, and hence the appliances of the temperance society are unequal to the emergency. This is not due to any lack of sincerity, of earnestness, or of correctness of principle or plan, on the part of temperance workers, but simply to a want of compatibility. There is an essential and radical lack of adaptability of measures to the evil to be corrected. In closing this chapter, it is but just to admit the impossibility of defining insanity in such a way that there need be no doubt or hesitancy in its diagnosis, and equally just to claim, that when inebriety declares itself as a disease, there is no cause for doubt or hesitancy in pro-

nouncing it as such, and that in this respect there is a
clear difference between the two. If, therefore, it is
impossible, or even extremely difficult, to establish the
fact of insanity in many cases, it is evident that the
boundary line between it and a perfectly sane and
normal and mental state, is beyond the limit of human
skill or judgment. We cannot, however, go far astray
in adopting the conclusion, that insanity and inebriety,
though allied disorders in some of their respective
features, are nevertheless distinctly different in other
delineations of their etiology and symptoms. I have
dwelt upon this topic at this length, on account of the
practical importance of the subject in its relation to
jurisprudence, especially, and, indeed, in relation to all
measures, both private and public, that have to do with
the prevention or cure of the evil.

Quotations from different authors, but chiefly from
the testimony of eminent alienists before the Parliamentary Commitee of Great Britain, in 1872, in relation to
the subject of this chapter.

Dr. J. Critchton Browne, at that time Superintendent
of the Lunatic Asylum at Wakefield, England, says:—

"I have come to recognize four forms of mental
disease, as being specially connected with intemperance.

"1. Mania-a-potu, or alcoholic mania, lasting from one
to two months. A genuine attack of mania, which is
always characterized by delusions; it is, as it were, a
prolongation of delirium tremens, followed by a good

deal of depression, and also mental stupidity, indicating failure of brain power, subsequent to the excitement.

"2. The monomania of suspicion, which is a form of chronic mental derangement, in which, without excitement or muscular trembling, we have delusions of suspicion.

"3. Alcoholic dementia, or chronic alcoholism, characterized by failure of memory and power of judgment, with symptoms of partial paralysis, generally ending in death. A very fatal form of brain disease."

"4. Dipsomania, which consists of an irresistible craving, etc."

"I think that in about one-third of these cases of insanity which I found to be due to drunkenness, there was a hereditary tendency as a coöperative cause with the drunkenness, creating the predisposition. I think that the hereditary tendency from drunken parents does not always manifest itself in insanity, but frequently in idiocy and crime." "Dipsomania is sometimes the cause of drinking, and sometimes the product of it." * * * "The excessive drinking in many cases determines the insanity, to which they are at any rate predisposed."

Dr. Arthur Mitchell, Commissioner in Lunacy for Scotland, says :—

"In a fit of ordinary intoxication we have really an epitome of an attack of mania, and a man who gets repeatedly drunk passes through *short attacks* of mental disease, which may eventually result in permanent cerebral disorder."

"There is mania-a-potu which is not intoxication, but is a maniacal excitement which comes on as the intoxication passes off."

"An attack of insanity of that kind may occur in a man who was never drunk before, and who was never drunk afterward." * * * "The man who frequently drinks and does not get intoxicated, leads up to insanity."

"The children of habitual drunkards are in a larger proportion liable to the ordinary forms of acquired insanity, or that insanity which comes on in later life. Many habitual drunkards are also strongly predisposed to insanity, and what they transmit to their children is really that predisposition to insanity which they have themselves."

Dr. John Nugent, Inspector-General of Lunatics for Ireland, says:—

"I knew an instance of a professional man, who at one time had been in respectable practice in Ireland; he married; he was unfortunate in his business, and as is too frequently the case, took to drink; he had four children, and each of the four children was either affected with insanity or was malformed; his wife also took to drinking, and she died in a lunatic asylum, and he, it was said, committed suicide." "I think drinking and insanity both act upon each other as cause and effect. I think that if there is a predisposition to insanity in an individual, indulgence in drink is sure to develop it; and, on the other hand, I think there are persons who show their insanity by a disposition to drink." * * *

" I have in my eye the case of a gentleman of fortune; the father of this gentleman was insane, and he had a brother; one of the brothers took to drink, and the other became insane *without ever indulging at all in drink; so that the hereditary disposition showed itself in one by actual insanity, and in the other by drinking.*"

Dr. Forbes Winslow, London, says :—

" In the upper classes of society, the insanity which can be clearly traced to habits of intemperance, of course, is not so great as in the lower stratum of society." * * * " There is a morbid craving for stimulants, which is clearly traceable to a brain condition; it is a form of insanity, although it is not recognized by law." * * * " You will never diminish the amount of pauper insanity until you deal with the great question of alcohol, and by legislation prohibit, as far as you can, its improper sale."

Dr. Robert Boyd, F.R.C.P., London, and University of Edinburgh, says:—

" Of admissions of drunkards to the Infirmary at Marylebone, in sixteen there was known to be an hereditary predisposition to insanity." " Of the sixty-three cases examined which died in the Somerset Lunatic Asylum, the majority had disease of the brain *prior to their continuous habits of drink.*" " I should think in the majority of cases, perhaps, the drinking habits produce the insanity, but in a great number of cases, symptoms of insanity precede the drinking."

ASYLUMS FOR INSANE AND FOR INEBRIATES.

INSANE ASYLUMS.

Some years ago I addressed a letter to several of the most distinguished physicians in charge of hospitals for insane, asking whether, in their judgment, it was proper to associate inebriates and insane persons in the same building, for treatment. The answers were all prompt and decided, in the negative. Inquiry was was also made as to the result of their experience in the recovery of inebriates who had been placed under the discipline of their asylums. The answers were equally decided, that the results were altogether unsatisfactory. This, I believe, is the universal testimony of all officers of insane asylums, as will be presently confirmed by resolutions passed at a recent annual meeting of the Association of Superintendents of such institutions.

In confirmation of this testimony of American physicians, I offer a few opinions from abroad :—

" I am more than ever convinced that there should be separate arrangements, separate asylums, and separate treatments."—*Dr. Alexander Peddie, Royal College of Physicians, Edinburgh.*

The Association of Dipsomaniacs with the ordinary inmates of an asylum operates injuriously in a twofold way, both upon the inebriate himself and upon his insane fellow inmates."—*Dr. Lauder Lindsay, Institution for Insane, Perth, Scotland.*

" I should object to mixing them. I think it would have a deteriorating effect upon those who were *not insane.*"—*Dr. Thomas Beath Christie, Superintendent of Insane Officers and Soldiers of Indian Army, at Ealing.*

The late lamented Dr. Forbes Winslow, so well and honorably known the world over, offers his protest against the practice of consigning inebriates to insane asylums. He says : " I went down to see a nobleman, not very long ago, who had been in a state of intoxication for four or five weeks ; he had not been sober during that time for one day, but I could see nothing in his mental or physical condition to justify me in placing him in a lunatic asylum. I know numbers of ladies, moving in very good society, who are never sober, and are often brought home by the police drunk. They are wives of men in a very high social position. I have been often consulted about these cases ; my hands are tied ; I could not legally consign them to the asylum." " If we have institutions distinct and apart from ordinary lunatic asylums, under a distinct course of direction, and perhaps, with a distinct class of inspectors and directors, they would, I think, tend very much to diminish the amount of drunken insanity."

Much additional evidence of the same character might be given, but I will offer only a few extracts, which are from the testimony of the late Dr. Donald Dalyrumple, M. P. for Bath, whose efforts gave the first practical impulse to British thought in this new departure. He reports to the House of Commons Committee, the result of his investigations in America on this subject, thus :—

"I find, moreover, that it is the custom of many lunatic asylums to receive inebriate patients *without certificates*, and *consequently, illegally*, and that a considerable revenue is derived from this class of patients, who are nearly all of the affluent class." He quotes also from the venerable Dr. Stuart, of Baltimore, who says : " In the State of Maryland we have the power to retain a man *non compos mentis* from drink, for six months ; but it is seldom done, because asylums for the insane are not places for drunkards. I am of opinion, that in inebriate reformatories there will be many a disheartening failure, many will rise, and fall again, but if help be afforded soon enough and long enough, many an object will be able to rise and remain permanently erect."

Dr. Dalyrumple further states, "I have made many inquiries of persons, other than those connected with inebriate institutions (medical, legal, clerical and lay), as to the effect produced by them, and received many and various opinions, but, on the whole, largely in favor of their utility."

In view of testimony from such distinguished sources, it is submitted, that the subject is not without grave interest to the common citizen, as well as to every one whose profession brings him in contact with either insane or inebriate patients.

The aggregation of numbers of all classes of insane persons in one building, in which the immediate charge of them must be delegated to ignorant ward-masters, is of itself sufficiently at variance with the advanced views of modern scientific thought, without increasing the roll of inmates, by including those who are not insane, and who are known not to be so, by the officers who receive and register them.

There seems to be a singular propensity on the part of some physicians, to magnify the existence of insanity by adding the vices and diseases of the same to their enumeration. Inebriates are mustered in, to swell the catalogue, while it is no more appropriate to include them in the list, than it would be to include merely queer and eccentric, or nervous and hysterical people. Numerous cases have come within my own knowledge, where inebriates, whom nobody would charge with insanity, have been committed to insane asylums, and detained in them. They appear on the records as insane patients, and may be admitted several times the same year, and be counted as many times as separate cases. Thus, an inebriate may go three or more times in the course of his life, to an insane asylum, and appear in the annual report as three or more distinct cases of insanity, and by such process, the public are led to believe there is a large increase in the number of insane. It may sometimes appear to families and friends, necessary to resort to such asylums with sudden cases of acute alcoholism, in the absence of any less objectionable place of detention ; and there seems to be no good reason why, under such circumstances, the cases so entered, should not be temporarily received as inebriates, if no other temporary shelter is within reach, instead of appearing in the published reports as so many insane persons. There are, as yet, comparatively few asylums for the special care of inebriates, and in the wide extent of country that must often be traversed to reach them, it is not surprising that both private physicians and patients, find it more convenient to avail themselves of the nearest lunatic asylum; but it cannot be just to the patients, to their relations, to the law, or to the community, to swell the record of insanity, by the indiscriminate addition of the inebriate class. Especially is it unjust, in view of the fact that the physicians in charge of the insane, almost unanimously agree, that inebriates are unsuitable and unsatisfactory inmates for their asylums, and that they themselves incline to discredit the teaching that alcoholism is a disease, and hence not a subject for hospital treatment at all.

Dr. Dalyrumple testifies before the British Committee that he conversed with two deservedly eminent specialists in insanity, whom he met in Philadelphia, on the subject of inebriate asylums, and that they threw grave doubts upon the permanency and reality of many of the so-called cures; and said that "those connected with inebriate asylums dealt in general assertions and flourishes which diminished greatly, when closely examined," etc.

If those gentlemen had made themselves familiar with the records of inebriate asylums, they would have discovered that a considerable number of the inmates had previously been in lunatic asylums under their own treatment, and had been discharged. As I write this page, I glance at such a record now before me, and running over the names of fifty persons admitted, as they occur consecutively in the beginning of a single year, twenty of them had formerly been inmates of insane asylums, and yet not one of them was found to be insane, either by the law or the specialists who received and discharged them.

The record might be pursued further, with an accumulation of similar evidence. The fact is, that there is a class of alcohol patients who are incurable, and who go from one institution to another, testing the qualities of each, and who take as much pride and pleasure in rehearsing their various experiences, as a tourist on his return from travel does in narrating the incidents and experiences of his journey. Such facts should inspire gentlemen who have charge of institutions, with a degree of candor, more than was exhibited by those experts in insanity to whom Dr. Dalyrumple refers, who became the unauthorized and voluntary exponents of the character and results of inebriate asylums in America, which they probably had never visited, and with which certainly they have had no practical experience. Dr. Dalyrumple, however, determined to prosecute the inquiries himself, and in doing so, by personal visitation, and careful study of the whole subject, he concludes and reports, that the managers of inebriate asylums on whose soberness and fairness most reliance can be placed, are candid enough to admit the defects and abuses to which their institutions are necessarily subject, and that notwithstanding these, the results are abundantly satisfactory and encouraging.

In dwelling upon this branch of the asylum view, I am impelled by a conviction that the people have been

strangely misled in this matter, and that the reactionary
impulse, which has now taken definite form, has produced an attempt at a change of base among superintendents of insane asylums, which has found expression
in the form of resolutions, adopted at their meeting in
Auburn, in June, 1875, as follows:—

"*Resolved*, That in the opinion of the Association of
Medical Superintendents of American Institutions for
the Insane, it is the duty of each of the United States,
and of each of the Provinces of the Dominion, to establish and maintain a State or public institution for the
custody and treatment of inebriates, on substantially
the same footing in respect to organization and support as that upon which the generality of State and
Provincial institutions for the insane are organized and
supported.

"*Resolved*, That as, in the opinion of this Association,
any system of management of institutions for inebriates
under which the duration of the residence of their
inmates and the character of the treatment to which
they are subjected is voluntary on their part, must, in
most cases, prove entirely futile, if not worse than
useless. There should be in every State and Province
such positive constitutional provision and statutory enactments, as will in every case of assumed inebriety, secure a
careful inquisition into the question of drunkenness and
fitness for the restraint and treatment of an institution
for inebriates, and such a manner and length of treatment as will render total abstinence from alcoholic or
other hurtful stimulants during such treatment abso-

lutely certain, and present the best prospects of cure or reform of which each case is susceptible.

" *Resolved*, Further, that the treatment in institutions for the insane of dipsomaniacs, or persons whose only obvious mental disorder is the excessive use of alcoholic or other stimulants, and the immediate effect of such excess, is exceedingly prejudicial to the welfare of those inmates for whose benefit such institutions are established and maintained, and should be discontinued just as soon as other separate provision can be made for the inebriates."

It will be proper, in view of such statements, to consider the subject from the standpoint of

INEBRIATE ASYLUMS.

In an official report made by a competent committee, to the "American Association for the Cure of Inebriates," in October, 1872, it is stated, that up to a year previous to the date of the report (that is, October, 1871), there were 5959 persons admitted to the several Asylums represented in the Association, and that of this number 5557, or 94 per cent., were voluntary patients. It is also stated by competent authority, that at least 33 per cent. of this number, have been cured. But the question is often asked as to their permanent cure. The answer is, that such a thing as a permanent cure of any disease, a permanent reformation from any vice, or a permanent conversion from a life of sin, to a religious life, cannot be honestly promised beforehand, or an-

9

nounced afterward, by any physician with his patients, any philanthropist with his vicious wards, or any minister with his lost sheep. The word is a misnomer in this intended application of it, but it has become the fashion among chronic objectors to the asylum treatment of drunkards, to use it, and it is repeated here only for the purpose of disclosing its absurdity, by presenting it in contrast with other diseases and vices, as follows:—

Of how many cases of insanity, when they leave institutions, can it be said their cure is permanent; that there is no possibility of a relapse?

How many criminals who are dismissed from the hands of justice, can be said to be reformed beyond the possibility of future failure?

How many converts to religion, are so permanently established that they cannot fall?

It is not in the power of man, safely to assert any such result of his own finite work. It may be stated, however, as a general fact which challenges scrutiny, that the cures of intemperance may be as sure and reliable as any other forms of vice or disease that present equally acute and complicated symptoms.

Inebriate asylums have demonstrated a few facts, at least, which cannot be gainsayed. Many intemperate men, who have entered them voluntarily, and conformed to their teachings, have gone forth to the world, stronger and better than before, and are still pursuing sober and useful lives, in at least the proportion above stated. One man out of three has been saved, and this against strong adverse circumstances, in most cases. It has been shown, also, that there are not a few cases of incurable alcoholism, which may remain quietly and soberly within institutions, for years together, and thus shield themselves from the risk of debauch, and their families from annoyance and danger. It is also proved, that asylums are a constant public rebuke and warning to the people on the subject, which has a deterrent influence in favor of temperance.

There are many persons who have been inmates of such institutions, who are among our most valuable citizens, and who, from the very fact of having voluntarily made public confession of their infirmity, by seeking asylum treatment, and equally public confession of their recovery, do not intend to falsify either the fact of sincerity in making the effort, or of earnestness in pursuing their sobriety, by any inconsistency in this regard, if it can be avoided.

There are, on the other hand, many who are professional debauchees, whose other vices are covered under the more visible fact of drunkenness, and who are more suited to corrective institutions, than to insane or inebriate asylums. Their chief purpose is self indulgence. They are constitutionally vicious and sensuous, and care but little for anything, that does not pander to the gratification of a low nature.

Such are not inebriates in the sense that is recognized by those who have given the most thought to this subject; and as elsewhere indicated, this discrimination should be recognized and maintained, by all who have to do with such persons, either in the departments of medicine, jurisprudence or morals.

In this connection, attention is invited to the following extracts from an address of Dr. Willard Parker, of New York, before the American Association for the Cure of Inebriates, in 1870, as showing another good that is accomplished by asylums. He says: "We have stated that the inebriate asylum is a school in which drunkenness is studied and treated. It is now proved to be a disease, and curable. We have learned that there are different classes of patients, whose condition varies, like their family history. There is one class composed of those who had healthy and temperate parents, and who have had the advantages of education. They have commenced drinking socially, and have indulged more and more frequently, until disease, as manifested by the depraved appetite, is established in the system. This class incurred guilt, or in other words, sinned, in the beginning, by violating the laws of the system, just as the over-eater sins against his stomach, and suffers from dyspepsia; or the over-worker sins against his brain, and induces insanity or paralysis. A large proportion of this class can be cured at an asylum, and the time required for that cure will depend upon the duration of the disease, and the amount of organic lesion which exists.

The second class is made up of those who are descended from a drinking stock; they have inherited a tendency or predisposition; have less guilt to answer for than the first class; are less curable, or if apparently cured, are more in danger of relapse. With this class an irresistible craving occurs in paroxysms, and if they can be shielded, for the time, from the means of indulgence, they are safe until the occurrence of another paroxysm.

There are nations or whole communities with whom this fearful tendency to drink is an inheritance, as we have seen, to the perversion of their whole character.

The third class is composed mostly of young persons who are depraved in all their instincts, and who do not desire either reformation or improvement. They are not subjects for the ordinary asylums, and in time wise legislation will cause provision to be made for them, that the community may be protected against their irresponsibility and lawlessness."

Through the agency of inebriate asylums, there has not only been a more critical study of inebriety, but a more clear and satisfactory distinction in the varieties of its forms. The aspects of vice, crime and disease have been more closely observed, and more clearly defined, and while in their corporate capacity they have kept clear of political and ethical discussions of the subject, they have done much to clarify the medium of observation in these respects for the people at large. More valuable additions have been made to the literature of the subject during the past few years, than for a generation preceding, and it has reached a class of minds that has not hitherto been inclined to the ordinary views of the subject.

That inebriate asylums can improve in their methods if they would realize their highest ideal, is admitted, but it is asserted, without fear of contradiction, that new and imperfect as they are said to be, they have accomplished larger and more practical results with this class of subjects than any system within the knowledge of the age.

It is well known that there are some drunkards who "recover naturally," that is, of their own unaided efforts. They "work out their own salvation" in this matter, and are among the heroic men of the times.

It is said that they constitute about three per cent. of the inebriate class; about ten per cent. of reformations are claimed by temperance societies, and it seems to me that they are entitled to this award, in addition to the quiet family work that is being done, under their influence, toward prevention.

These facts, taken together with the fact that inebriate asylums, homes and reformatories record at least thirty-three per cent. of their cases restored, and that the public sentiment is strengthening every day in favor of sobriety and virtue, there should be no cause for discouragement in any quarter, nor for any other rivalry between the different methods, than that which is born of high purpose, and earnest effort to accomplish the most good for the individual, and the general public. In the establishment of asylums and homes for inebriates, two classes of effort have been recognized, and the practical results have satisfied those who have labored respectively in each, that great good may be accomplished by both.

The homes or reformatories, rely upon the moral and religious means which constitute the elements of all purely Christian work, and place the physically remedial measures, in a subordinate relation to the subject.

The latter are generally located in cities, and do noble service in organizing the inmates into clubs and societies, the membership in which, is not necessarily lost by discharge from the homes.

It is the privilege of such persons to continue their membership, attend the weekly meetings, and aid each other in the outside conflicts of life. Many men by such a course, recover their manhood and become useful citizens, and thus unwittingly, but nevertheless truly and forcibly, reprove the notion that legal inquisitions, commitments by courts, compulsory restraint and protracted confinement, are necessary for a cure in such cases.

Asylums, which recognize more fully the disease theory than is done by the homes, and which generally receive the more chronic cases and those known as dipsomaniacs, study their cases physically as well as morally, organize them for the cultivation of the æsthetic qualities of their nature, instruct them as to the higher duties of life, encourage the largest *possibilities of their being*, nurture their hopes for a better future, and without undue and degrading "compulsory confinement," release many, after different periods of residence, who go into the world again and become useful citizens.

Such, I trust, is a fair representation of the principles and outline of plans, adopted by the institutions in the United States for the cure of inebriates. Justice may not, however, be done to the subject without reference to the laws which provide for arrest and commitment. It frequently happens that men run recklessly into excess, damage their estates, expose their families to the risk of want, and are themselves dangerous to be at large when intoxicated. The law provides that in such cases, arrest may be made of the offending party, on complaint of a near relative; and after a hearing he may, at the discretion of the court, be committed to an inebriate asylum for a term of months, a year, or even longer, his property being placed in charge of a trustee, for the benefit of his family. The law thus serves the purpose of defending the family from danger, economizing expenditure, and allowing the inebriate to come to himself, meditate upon his condition, and if possible, reach a state of mind in which he may be capable of deliberate action, and resolve to avail himself of every help to a better course of life.

So far as he is related to institutions, however, the treatment is the same as that of the voluntary patients. No difference in the management marks the relation he may hold towards his family, his property or the law, unless he himself exposes the facts, or refuses to accept the proffered confidence and freedom. Asylums and homes for inebriates cannot conform to the legal standard, so far as their treatment is concerned, though the law may be used to induce inebriates to conform to circumstances and conditions, which shall promote their own good. The genius of institutions, is to succor and restore the diseased and fallen. The genius of the law, is to administer justice, and protect the family and society. Institutions stand between the family and the drunkard; the law takes care of the one, and institutions of the other, the purpose of both being to accomplish the same end, though each operates in its own peculiar way.

Institutions need the law as a resource, which may in times of emergency aid them in their benevolent dealings with men in a state of unconsciousness. But the spirit of institutions appeals to such persons to rely upon other forces, than those which the law apprehends and declares, so soon as the intellect, and moral sense, are clearly within reach. The law needs special institutions for such cases, in order that the ends of justice, and the good of all parties may be secured, and thus the intention and operation of both, seems to be entirely beneficent.

The object of this chapter will have been reached, if it shall strengthen the sentiment, that institutions for the insane are unfit places for inebriates, and that those who conduct such establishments have suffered damage to themselves, and to the benevolent work in which they are engaged, by so long consenting to be custodians of a class of persons, who are not believed to be insane; and also if it shall exhibit the true nature, principles and results of institutions for inebriates.

THE INEBRIATES' VIEW.

We come now to consider a most interesting feature of the subject, as presented from the standpoint of inebriates themselves; and it is submitted that no class of persons are more directly interested in the subject, or have a better right to be heard, than they.

Surely the lessons of suffering, and the experience derived from it, are entitled to weight, in a discussion which comprehends the victims, who endure the suffering, and learn the lessons.

I have before me a communication which was carefully considered, and unanimously adopted, at a meeting of the inmates of an inebriate asylum, and duly forwarded to a convention of physicians and friends of such institutions, which was held in New York, on the 29th and 30th of November, 1870, from which the following extract is taken, to represent their side of the question.

"GENTLEMEN:—We are aware that in offering to you our views upon the grave subjects whose discussion has brought you together, we occupy the position of the condemned criminal, who, his case having been adjudicated, is simply, *pro forma*, asked what he may have to say, ere the already determined sentence be passed: and yet we trust, in appealing to you as our advocates, we have come to those whose careful examinations, enlarged knowledge, and generous motives, have enabled them to set aside hasty conclusions, and common prejudices; and that through you we may appeal again to the bar of public opinion, with the hope of a kinder hearing, and a revised judgment, which may perhaps be productive of higher good.

"In common life, so intimately mingled is the vice of intemperance with some of the offences of the professional criminal, that to most persons they are but synonyms. The one is but too often added to the crimes of the other, and appearing, as they do, thus yoked, in our courts and penitentiaries, it is hardly strange that even the good and virtuous should esteem them identical. It is not necessary that we should deny this with reference to ourselves; for neither our friends nor our worst enemies will make against us this charge.

"Doubtless, to ourselves, as well as to others, the cause of our condition is a mystery. We have all been educated with a deep respect for religious obligations, which we still retain, some of us having been church members. Some few have been accustomed to the use of alcoholic stimulants in our homes, and find ourselves victims to their power, while other members of our families, brought up under the same influences, have escaped unharmed, and are now occupying active positions in the busy world, still indulging more or less freely, and with apparent impunity, the appetite which has been *our* ruin.

"Some of us, in our early business life, were taught to believe that an open-handed liberality, and the free offer of the glass to our customers, was necessary to success. Others remember that in the pursuit of our professions, in the freedom and irresponsibility of a students' life, we

were surrounded by those who joined freely in the
convivialities of the drinking saloon and wine supper;
and now, as we look around, and ask for our quondam
companions, we find a few, and they perhaps the most
brilliant and beloved of our circle, conquered by our
common foe; but the large majority have thrown off
the wild habits of those days, and are now settled in
their various homes, in successful business.

"These are simple facts that startle us, as we recur to
our own unenviable situation, with the question, why?
Gentlemen, we do not attempt to answer. We ask of
you, our judges, to reply.

"Were they upon whom the tower of Siloam fell the
worst of criminals?

"It is not our intention here, however, to argue the
question of criminality. While we confess to our
full share of human weakness and sin, and acknowledge
our unfortunate dependence upon society and friends for
protection and relief, we have, nevertheless, an inalienable conviction of our right to share, in common with
others, the elevating influences of our Christian civilization. Has society extended to us this right? In
order fairly to answer this question, we respectfully
submit for your consideration the following propositions:

"1. That a social ostracism is practiced toward us,
which is not practiced toward other members of families
or society who have vices and diseases that are equally
offensive to morals and equally damaging to the community.

"2. That church ostracism, in many instances, deprives

us of the very sympathies and forces that should combine for our relief and restoration.

"3. That we suffer from legal disabilities, by which offences committed in a state of unconsciousness, from intoxication, are on this account punished with more severity; while the same offences committed during the unconsciousness resulting from insanity, or other diseases, are mitigated or excused, on account of the same.

"4. That our sorrows and sins are made texts for sermons; our symptoms and misfortunes are caricatured by lecturers and performers, and we are exposed alike to odium and ridicule, which has a most depressing and damaging effect upon our mental and moral nature, and directly predisposes to results against which we would guard.

"5. That we are expected to change or overcome our constitutional tendencies, and reform our lives, under a degree of pressure from all classes of the community, such as is brought to bear upon no other class of individuals.

"6. That in view of these facts we need places of refuge or asylums, where we may escape the depressing influences to which we have referred, and where, for a time, freed from the temptations and associations amidst which we have been led astray, we may regain that moral tone and power of will which can alone fit us for the duties and responsibilities of life."

The fact of "social ostracism" is presented, in view of the domestic training of youth, upon some of whom it seems to have no deleterious influence so far as drinking to excess is concerned, while upon others of the same household the calamity of intemperance falls with resistless power.

Is it any marvel that in families trained under the same roof the unfortunate victims of appetite should be surprised with their attitude in relation to those who have drank at the same table, followed the same parental example, and obeyed the same instructions with themselves, and yet who have avoided this form of excess, and risen to honorable and prosperous citizenship?

It is, indeed, no wonder that they are startled at their own "unenviable position," and that they cry out, "The cause of our condition is a mystery;" and that they repeat the inquiry, "Why is it?"

This inquiry, too, has peculiar emphasis, in view of the fact that others of the family circle "have vices and diseases that are equally offensive to morals, if not equally damaging to the community," who escape the ostracism, which is visited chiefly, if not entirely, upon the inebriate.

Of their "legal disabilities," they have less cause to complain. It is the province of law to dispense evenhanded justice. It knows no such thing as sympathy. It cannot compound with domestic affection, or social friendship. It must shield the family and home from outrage, as well as the inebriate from needless exposure and danger. It is inexorable, and it may sometimes be unwise.

There are no class of persons who are more self

deceived than the intemperate. Suspicious, capricious and apprehensive, it is natural that they should have distorted views of their surroundings and of the influences which they have to oppose, and yet the following language of the address just quoted may be accepted as true :—

" We are expected to change or overcome our constitutional tendencies and reform our lives, under a degree of pressure from all classes of community, such as is brought to bear upon no other class of individuals."

The truth is, society has not until recently begun to recognize the fact of " constitutional tendency," but has attributed the state of drunkenness to the heedlessness, if not to the willfulness of its victims.

The inherited tendency to insanity has been appreciated, and large appropriations of public money and munificent bequests of private wealth have been freely bestowed for the care and protection of the incurable of this class, and for the recovery of those who are supposed to be curable.

Indeed, public sentiment has become so softened by the presence of insanity, that a protest has been made against the confinement of the dangerous or homicidal insane, either in the same building with the more harmless, or yet in the penitentiary with sane convicts. So urgent has been the sentiment of philanthropy on this subject, that the blunder has been committed of adopting the misnomer of "criminal insane," as applied to such, and of proposing to erect separate institutions for their detention and confinement.

No enlightened person questions now the constitutional tendency to insanity, to deaf muteness, to blindness, to idiocy, to pulmonary consumption, to gout, to rheumatism, and many other forms of disease; and all admit that with proper hygienic regulations they may each be considerably modified, if not prevented. It is even now asserted by many wise jurists, that the tendency to crime is an inheritance, and numerous proofs of this fact are coming to light.

Inebriates, however, in the popular estimation, have been excluded from such considerations, and no allowance has been made for them on the ground of transmitted infirmity. It must be that they feel the abrasion which follows this cold and unsatisfying response to their weary, oppressed and appealing natures. Think of the secret struggles that have no recognition, the infirmities that have no strengthener, the bruises that have no healer, the palsied hopes that have no faith left to build upon, the losses of property, home and love, with no one to recompense, and the deep and biting remorse that reproaches and poisons the inmost soul with the reminiscences, first of weakness, then of sin, then of sorrow, shame and penury to self and family, and finally of despair !

"The drunkard endures more painful conflict to overcome his passion, in a single day, than many a passionless passive, soul, who knows nothing of such an appetite, suffers in all his life," has been said by some one. There can be no doubt that there is some truth in the declaration; and while I would not excuse the drunkard for willing and known violation of the laws of his being, I would invoke for him the same consideration on the part òf society, that is so generously expended towards others, whose sins may not be so public as his, but who are as certainly living in violation of both natural and revealed law.

No men have keener susceptibilities than those who are conscious that they have fallen.

None are more anxious to rise, and yet their very acuteness to wounds, prevents their rising.

I have known some to be indifferent to religious appeals, and to avoid the means of religious instruction, who have been restored to sobriety under the overpowering motive and influence of business interest, and from a decided and obstinate purpose to be independent of any purely moral influence; while those of deep sensibilities, who have been cultured in religious thought, and trained in religious life, from the very fact of the wound to their higher nature, and of the reproach they have brought upon their professions, and their former lives, sink into despondency, and, like lonely captives, give themselves up to perish of thirst, amid the wilds of desertion and penury.

There is a profound interest in this fact of the secret selfhood of the inebriate. It amounts to an exaggerated self-consciousness, which renders him the most difficult of all men to deal with, in matters which most concern him. At a certain stage of his disorder, he becomes an egoist of the most positive type. The very conduct of sober society towards him, drives him to himself, and

10

intensifies the consciousness of his personality. The more he feels himself drifting away from the domestic and civil amenities of life, by his indulgences, the more he welcomes the allurements of his own passion, becomes the keeper of his own thoughts, and the contriver of his own methods. These are symptoms of his dipsomaniac tendency. When numbers of such men meet in society, and form fellowships, the one thought possesses them. It is a secret laid away in the heart of each.

The inward conflict is between the impulse to satisfy the propensity of their nature, and the ever haunting visions of home and duty, that follow them, and of tongues that whisper, while conscience is full-voiced with warning. Still the conquering impulse drives, the will lies prostrate, and the debauch is accomplished.

When it is over, the horrors come; the terrible strife between the inner and the outer man.

The moral, for the time may get the mastery, and solemn resolves be made, resolves that are too hastily made, because the will and the purpose, which are largely dependent upon physical conditions and forces, are defective, and insufficient for the crisis.

These sketches of the "inebriates' view" are believed to be real, and it is an important question how far society has adapted its appliances to their condition.

I am aware that there is a reverse side to the picture. Society itself has its rights, and it must protect itself. The family has its rights, and they must be secured; but the question is, whether society and the family cannot be better served, by looking at and dealing with the inebriate as he is, and not as a false sentiment conceives him to be.

In the following chapter this question will be considered.

HOW TO DEAL WITH INEBRIATES.

As within the whole range of pathology, there is no more intricate inquiry than into the actual signs of the alcoholic diathesis, so there is nothing in the domain of the materia medica, more difficult to find than a remedy or remedies for it. Therapeutics also, the science of preparing and applying remedies to diseases, stands in the presence of inebriety, listless and strengthless, unless we recognize in its scope, the entire range of hygiene and dietetics, granting at the same time, a share of influence to measures that are effectual in awakening the moral and emotional nature, and relying upon its agency in arousing antagonisms to the disorder, by its reflex influence upon those nerve centres that are the most directly concerned in the development of primary symptoms. In this connection, also, a large place should be awarded to prophylactic methods, which, while they may not directly affect the present, must have an important bearing on the future of any given case, or its offspring.

We shall, therefore, commence the discussion of this topic with the assumption, in the very onset, that no single specific remedy for inebriety, has as yet been discovered, either in the realm of physical or mental forces.

As is the case with a great variety of neurotic disorders, the co-operation of the patient is essential to the remotest degree, to a proper management of the case. On his part, and on the part of his family, it is requisite that there should be an intelligent apprehension of the conditions to be overcome, and a confident reliance upon methods to be employed.

The first thing to do, is to change the standpoint of observation ; to get on a common ground and occupy it. Study the subject from a new outlook, recognizing the dogma of disease, as the basis of all effort toward prevention or cure.

Notwithstanding all that has been said and written to the contrary, facts demonstrate that inebriate asylums and reformatories, as they are severally called, have done more, in proportion to the number of inebriates who have been under their care, to restore and establish them in their normal relations to the family and to society, than any other instrumentality.

This has been accomplished, not by specific medication, nor yet by appeals to the moral sense. It has been done, not by pledges, nor yet by the enforcement of disciplinary regulations. It has been done at great odds, by the recognition, first, of a diseased body, and its dependence upon extrinsic agencies to fortify it against the morbific forces which have disturbed its normal equipoise. It has been done, secondly, by the creation of a model family bond, which recognizes mutuality of interest, and an obligation to maintain it, by the exercise of mutual confidence and trust. To build up the waste places that have been desolated by the existing morbid elements, which has rendered the case susceptible to the toxic touch of alcohol, has been the chief purpose of all interested parties. It has been done, thirdly, by the faithful observance of hygienic laws, and the avoidance, so far as possible, of exciting causes. There is no mystery, nor has there ever been, about institutions for the cure of inebriates, and so soon as any household in which this disease of inebriety becomes apparent, shall be equipped with the simple appliances and methods, that have been found to be essential to such asylums, and the genius that has inspired and controlled them, shall become the prevailing inspiration of the family, there shall have been inaugurated a revolution · in the life of that domestic circle, that shall make itself felt as a power in the social life of a community.

The first thing to do, then, is to place inebriety along side of its kindred disorders, by removing it primarily from the domain of morals, and dealing with it, as allied conditions under other names are dealt with. Everybody knows that there are diseases that are a source of constant trial and disappointment, both to families in which they are found, and to physicians who are called to treat them. Unclassified, and even unwritten disorders, nameless and mysterious, which are the outcome of our intense American life, for which little can be done, without time, patience, and earnest research into the laws of our being, and the multiplied forces about us, which tend to disturb them. From their very complexity, not always preventing the patient from attending to business, and yet creating distress and disorder, wakefulness at night, and irritability and restlessness during the day, they cause, not only discontent and unhappiness on the part of the invalid himself, but distrust and suspicion on the part of associates and friends.

While I write this paragraph, a lady calls for advice, with the following story : Aged 35; frequent and unaccountable palpitation of the heart ; obstinate constipation, requiring frequent medication ; gloominess, distrust and suspicion of others, making herself and her family unhappy. Among her peculiar and specific symptoms, are the following : Odors of flowers bring on paroxysms of short and difficult breathing (hallucination of the sense of smell); the act of deglutition is often accompanied by severe pain in the breast, and spasms of the larynx, so that she is obliged to leave the table ; with it all, she is apparently in good health ; sleeps and eats well, but is oppressed and moody, because she is not understood. Having but little sympathy from · others, she has become self-conscious to a marked degree. Her friends tell her that her sufferings are imaginary, and that all that is necessary for her to do, is to exercise her will, and subdue her morbid imagination. Hence her suspicions. She sometimes, in her forced confidence in others, mistrusts herself, and begins to doubt her own sincerity, until an actual pain shoots through her frame, or some part of it, and her heart repeats its throbbing with undue violence, and she is re-convinced that her sufferings are actual, and entirely beyond the control of her will.

At times, she has temporary local paralysis, now in one limb, and again, in another, and yet, in the intervals, walking and working with the alacrity of health. She is a mystery to herself, and feels that she is the cause of extra care, and a grievance at home, and then she desires to escape from it. There is no place so uncongenial as home. It is the scene of her conflicts and sufferings, of her suspicions and fears. The faces that she sees about her, are sometimes expressive of sympathy, but oftener of doubt and suspicion. The ministries that come to her during her hours of pain and struggle, she feels to be the ministries of necessity, rather than of cheerful love. Unhappiness is her allotment, and the tendency of her bruised and neglected (?) nature, is to employ narcotics that she may find relief, and yet, realizing the danger of such use, seeks counsel that she may escape the debauchery of drugs.

Such cases are of frequent occurrence. Their name is legion, for they are many. Among them inebriety has a conspicuous place, and it cannot be separated from others of its class, without doing violence to facts, and a grave injustice to the victims of such neurotic conditions.

Indeed, the case of the female just narrated, is similar in many respects to that of inebriates. She was on the way to intoxication. She stood on the brink of the precipice. She felt that she would be benefited by a narcotic. She desired to forget her distresses and herself. She sought not only solitude, but oblivion. Before taking the leap, however, she determined to make one

more effort, and find, if possible, the cause.   It required
no unusual discrimination to discover, and locate it.
It was of pelvic origin, and readily relieved.   With
local relief, came general comfort and domestic happiness.   It would be as sensible and logical to ask such
an one to sign a pledge, that she would never again
display her suspicions and fears, as it would be to ask
an inebriate who was subject to the same annoying
suspicions, and who suffered similar pains, frequently
induced by local causes, never again to be disconsolate,
moody, restless, suspicious, which are symptoms of his
inebriety, and but for which he would not seek relief
in the intoxicating bowl.

It is said that the propensity to drink is not
induced by physical causes.   Read, not only elsewhere
in these pages for the most authoritative testimony to
demonstrate the contrary fact, but the following, from
my friend, Dr. Crothers.   (See *Medical Record*, Nov.
4th, 1882.)

" A strong, vigorous merchant, with no heredity, and
temperate, suffered from a partial sunstroke.   He
remained, greatly debilitated, for two months, in bed,
and then began to use spirits to excess, and was a continuous inebriate up to death, four years later.   He
made great exertions to recover, by the pledge and
prayer, but failed, and died of dementia.   The drink
craving was clearly traced to the brain injury from
sunstroke."

Another :—

" B., a vigorous man, temperate and correct in all

his habits. At 31 years of age he married, and his wife was killed on the wedding tour, in an accident." Profound grief followed, with inability to attend properly to business. He became sleepless, with loss of appetite, which resulted in drunkenness and death. The sudden shock, technically a " psychical traumatism," was followed by a change in the normal functions of the nerve centres, which was the starting point of a general moral and physical degeneration. " He talked and reasoned clearly, and made efforts to recover, signed the pledge, asked the prayers of the church," etc., but the change of brain and nerve integrity was the beginning of a general disorganization, and things went on from bad to worse, in spite of all effort by himself and friends.

"Financial disaster came upon another, a wealthy merchant of much character, exemplary and honorable, with an inherent dislike for the taste and smell of spirits, which were never allowed in his family. Sudden poverty, the loss of a wife, and a scattered family, all within a few short months, turned his hair gray and left other marks of physical change. Suffering told upon him, and the changes that were wrought in his constitution, were in the direction of dishonesty, untruthfulness and intoxication. Great efforts were made to save him, which he appreciated and tried to assist, but in vain."

The same thing we see as the result of acute disease or accident;—change of character, perverted tastes, leading to vice and immorality. We need but walk the wards of a lunatic asylum, to hear the most obscene utterances from the lips of virtuous and refined women, the most profane and boisterous language, from men who, when sane, were renowned for their purity of life, and honorable intercourse with their fellows.

Insanity and inebriety come from the same stock, though varying somewhat in feature and career; and similar injuries to nerve structure, or function, will be developed alike, both in the insane and intoxicated.

This rather lengthy divergence from the line of thought that was being pursued, seems allowable for the purpose of emphasizing the fact, that inebriety may be classified with other neuroses, and should be so considered, in order to enter upon a proper course of treatment.

The new departure for patient, family and physician, involves the acceptance of the doctrine of disease, as it has been disclosed in these chapters. If disease, there need be no shame, or self reproach, attached to the fact of its existence, unless, indeed, it has been developed by a reckless and wanton life. Neither should there be reproach or rebukes on the part of others. The gulf between sober and intemperate people, need not be the dark and dismal chasm it now is. It may be bridged over by a strong and steady span, that will allow both to meet, and occupy a common ground. There need be no concealment of symptoms which preface a debauch, nor of avoiding the issue that is sure to come, if premonitions are disregarded. I have so frequently seen cases in which an approaching debauch has been prevented, that I dare speak with assurance upon the subject. At this moment I have a vivid recollection of an intelligent physician, who at times was subject to attacks of mental depression, which were invariably followed by a drunken carouse. He came home one evening in great haste, showing unusual evidences of excitement and apprehension, feeling, as he did, that he was drifting into drunkenness, and urgently pleading with me to save him from such a calamity. A brief examination revealed a single fact, upon which the whole case turned. He was constipated, and an active cathartic cut short the paroxysm, though he requested to be guarded or restrained till the medicine should take effect. Had he concealed this fact, or not been made aware of its reflex influence upon his nervous system, he certainly would have fallen into the condition he was anxious to avoid. Finding relief, and learning from this single experience, how easily he could be thrown from his balance, he was always afterward on the alert, and able to discern the premonition, in time to prevent an attack.

The premonition in his case, was not always the same. Constipation it might be again, or headache, or indigestion, or neuralgia, but whatever it might be, he had learned to look upon it as an exciting cause, or at least as a friendly premonition, and in using suitable remedies for the relief of his symptoms, his balance was maintained, and a sober life was afterwards the result.

Inebriates, however, are not, as a rule, persons who begin a career of intemperance by looking out for premonitory symptoms, or who try to learn the grave lessons which they teach, with such accuracy and emphasis.

They are, however, persons who, once aroused to the
fact that their inebriety is due to physical causes, which
it is often in their power to discover, and sometimes to
prevent, may be strengthened by such knowledge,
and so far fortified with the means of self help. Their
most intimate kindred and friends should likewise be
informed, that though the objects of their concern may
sometimes be overpowered by such symptoms, and fall
into degrading excess, all their inner moral life should
not therefore be assailed, as debased and ruined, but
that they should stand in the presence of home and
of society, as conquered victims of a physical disorder,
for which aid should be sought, as in other bodily ailments or infirmities. Could society bring itself to such
a standpoint, or could the inebriate himself, feel that
his own consciousness of physical defect or injury, was
sustained by the intelligent sentiment of the society in
which he moves, he would naturally and promptly seek
counsel and relief, without the sense of shame and dis-
honor, which is now the millstone about his neck, that'
sinks and drowns him and his comrades, in hopelessness
and death. In this connection, it should be stated that
these prodromic symptoms are more apparent, and more
accessible, in that class of inebriates who are known
as periodical, or more properly, paroxysmal drinkers;
who are sober weeks or months, without any inclination
to indulgence in the interval. The case of the neurotic
female, just narrated, represents a class of symptoms that
should be looked for in such cases.

Common expressions among such persons, are, " I had

the blues." "I was down in the mouth." "I was cross and ugly at home, and suspicious." "My appetite failed me." "I was restless, sleepless, had night sweats, and was generally out of sorts." These are common modes of expression, but they are definite and descriptive, and important also. It remains for the adviser, medical or family, to ascertain the cause of this condition, in its several phases, and then to remove or modify it if possible. If this can be done, the debauch is prevented, or even arrested after it has commenced. This has not unfrequently happened in my own experience.

The habitual drunkard belongs to another class. He has no protracted intervals of sobriety. He frequents the saloons before breakfast, perhaps, and keeps up his potations till midnight. He does not *go off* on "sprees," as the term goes. He is not free from the alcoholic impression at any time. He is more or less alcoholized continually. Can the habit be broken? The moral sense of such persons is frequently blunted, it is true, and yet there is no occasion for despair. The will power is enfeebled, it is true, and yet if the will of another is permitted, for the time being, to assume control, there is hope. How feeble the will is, how powerless sometimes, is witnessed in the frequent violation of pledges and vows. It is said by some, that the will is always supreme, and that it is always at command, and only requires to be exercised in the right direction, to secure desired results. Facts do not sustain this view. The poison of typhoid fever prostrates a strong man, and he becomes as a child. Why not the poison of Alcohol?

The wear and tear of life exhausts nervous energy, prostrates vigorous manhood, and subjugates the will. The unnatural friction of a dissipated life, with all the depleting forces of excess and debauchery, will do the same. The will-power of an alcoholized person becomes a toy in the grasp of an overpowering passion, that is for a time invincible, and is only subdued by the expenditure of its own violence.

To place the constant drinker in circumstances that are favorable to his recovery, he should be under the control of a ruling mind, and give his consent to such an arrangement. He naturally dreads the immediate withdrawal of his accustomed stimulants, and the sense of fear of the consequences of such withdrawal, is at once an obstacle, and a hindrance. My practice has been, in treating hundreds of such men, to assure the timid, trembling, fearful cases, that they shall not be allowed to suffer, or subjected to any risk, by a too sudden deprivation of all stimuli. They shall still be allowed, if necessary to sustain them, a certain quantity, medici-' nally, the difference being, that instead of taking it at their pleasure, for the purpose of exhilarating them, I shall administer it according to my judgment, to prevent their sinking, or falling to a dangerous level.

This is a change of base at once. It is reasonable and consonant with sound practice in other maladies, and without a cell and a key, some such argument is especially wise. It often assures confidence. To avoid shock in such cases, is as important as it is in surgical practice. Habitual inebriates are not usually men with

a strong inherited diathesis, similar to that of the paroxysmal drunkard. They have possibly developed the
disease by the inordinate use of alcoholics, and in so
doing, have set up a series of morbid changes, which are
symptomatized by a variety of functional disturbances,
that require close watching and careful management.
Some form of liver or kidney disease is almost sure to
attach to the history of such cases, or a complication of
some kind, is frequently noticed, as having existed prior to
the adoption of the habit. In view of the general disturbance of function, and of possible structural change,
it is safer to proceed cautiously and work patiently.

Nothing can be more unphilosophical or unreasonable
than to expect such a person, by a mere act of will, or
by subscribing to a pledge, to abandon at once the habit
of years. There are disordered functions to be corrected.
They may be cardiac, gastric, renal or hepatic, or any
combination of these, and the natural order of things
will not be restored by any mere act of the judgment or
will, nor yet by the punitive behests of law.

First *interrupt* the habit, cut off the allowance at once, if
it can be done without risk, substitute such remedies as
the symptoms may indicate, and the improved capacity of
the system will allow. The reason why so many fail to
keep the pledge, is found in the fact that they take it
when, in the very nature of things, it is impossible for
them to live by it. During the existence of these functional disturbances, which are continually aborting every
effort of the mind to control itself, it is folly to attempt
the assertion of a normal self. Much more is it absurd

to rely upon any such proceeding, to arrest or control
organic deterioration; and the use or disuse of drugs,
whether alcoholic or not, depends largely upon the
indications presented by the organic lesion whatever it
may be. By pursuing such a course, many persons who
have been addicted to alcoholic use in excess for many
years, have been restored to a state of sobriety, though
they are frequently, perhaps generally, left with an
entail of chronic disease, which finally carries them off.
They die from chronic alcoholism after years of total
abstinence. They may appear on the mortuary list as
having died from paralysis, brain or liver softening, or
other form of vital impairment, but the true pathological description of cause, would be *alcoholism.* There
are, doubtless, living to-day, many men who were once
intemperate, and who will never again use alcoholic
beverages, who will die of alcoholism. The tissues
have been poisoned, and so continue through years of
sober living, in which they are said to have reformed.
They have abandoned the habit of drinking, and thus·
given nature and remedies a chance to do their part
toward reinstating the patient in a normal relation to
society and to the world. The machinery, however, is
impaired beyond renovation, and it only works with
the semblance of normality, by the strictest watchfulness
and care.

An excellent gentleman once called on me in behalf
of a ward, for whose welfare he was deeply concerned.
The young man was an habitual drunkard, though the
proprietor of a temperance restaurant, in which employ-

ment he engaged as a means of keeping him in the path of sobriety.   After describing the case and expressing a strong desire that I should visit his ward, he made this remark : " I do not expect you to cure him, Doctor, for he is always under the influence of liquor, and yet able to attend to his business, but is restless, excitable, busy, and all the time on the go.   He does not stop long enough to think, and I have lost hope of any human agency being of use to him.   The most that I can expect you to do for him, is to get him sober, and keep him sober, and quiet long enough *to give God a chance at him*.   That is all."   This proved to be an interesting case.   He was sobered, became quiet and thoughtful, and doubtless the coveted "chance" was secured.   He was a youth of abundant resources, fond of history, a poet, and of considerable literary culture.   With such resources within himself, and by the use of suitable remedies to meet the unusual conditions of his body, his course of life was changed for a time, so that he was really called a sober man ; but disease had fastened upon him, and he could not for any considerable length of time endure the struggle it cost him to keep sober, and finally, as I believed, died from alcoholism, though the record of the Health Officer after his name was, "Died from Paresis."

Such is the history of many an habitual inebriate. They struggle with themselves, by themselves, and against themselves, and succeed in walking soberly for a time ; but the poison has written its own name on their vitals, and whatever the death certificate may record, the fact of pathology is, "Alcoholism."

11

Thus far we have been considering symptoms, most of which were preceded by prodromic signs, that announce their approach so regularly, that they are finally recognized as belonging to the condition which it is our object to overcome; but the truth is that this is not always the case.

It happens in the experience of most families who are visited by this calamity, that drunkenness forces itself upon them, without being heralded beforehand, by any observed admonitory symptoms. It is only by experience and repeated observation, that they are understood, and their approach looked for. It is after repeated surprises, that the intelligence is thoroughly aroused to anticipate them, but that they usually exist is true. They may be overlooked by the patient himself, or if he is warned by them, he fails to recognize their correlative relation. Such may be called cases of acute intoxication, and how to meet them is a difficult problem. A few suggestions, however, may be appropriate. Some inebriates are boisterous and threatening; they come to their homes infuriated by the recollection of some previous, and perhaps long-buried family discontent, in which they invariably esteem themselves to be the aggrieved party, and pour forth their anger and abuse toward those whom they should protect. In such a state of things, opposition, reproof, excitement of manner or utterance, or the exhibition of timidity, increases the difficulty and risk. On the other hand, too much kindness or strained attention, with a view of overcoming his rage by kindness, he intuitively regards as unnecessary and undeserved, and it thus is frequently the occasion of increased excitement. It is better to be quiet and not retaliative; indifferent, rather than too regardful of his demands or threats; and when addressing him at all, to do so with a calm, deliberate, self-contained manner; and generally in a low tone of voice, in which, however, there should be the evidence of firmness and truth. I have frequently quieted the most excited and boisterous, by a calm semi-whisper, communicated in a confidential style, that almost invariably commands attention and respect. Even in delirium tremens, or in that prodromic wavering of intelligence and paroxysmal shuddering, which are its precursors, a soft answer, in a measured whisper, in a close, confidential manner, will often soothe the patient, and render him amenable to treatment that would otherwise be resisted. Especially if the key-note is struck of some favorite theme, on which his mind is known often to linger, it is surprising to witness, how prompt is the transition to a state of comparative composure.

On the other hand, there are cases that can be most effectually controlled by firmness, and a quick decisive manner. I have known some men, intellectual inferiors to their wives, stagger to their homes, with scarcely a consciousness of their going, blunder across the threshhold, and come suddenly in the presence of their wives, who would scarcely stop from their occupations longer than to order them immediately to their rooms and their beds. And no sooner would the order be given, than it would be obeyed, with all the submission of a truant child returning with penitent steps to his home.

The physical treatment in such cases is simple. Nourishment, in a concentrated form, is indispensable, and nothing is better, or more likely to agree with the stomach, than milk, or well-seasoned beef tea. Sleep must be induced, and if the stomach is able to retain the nourishment, sleep will generally follow.* If, however, there is continued insomnia and a probability of a return of excitement, as there will be in such circumstances, Hydrate of Chloral is the most decided and effective medicine to be used. Though I have frequently found that a moderate inhalation of washed sulphuric ether, will produce a temporary calm, which, if frequently repeated, with the return of a tendency to wakefulness, will eventually calm the nervous irregularity, and allow good, wholesome, refreshing sleep to come. The question here arises, is alcohol in any form, admissible in the treatment of these acute cases. I am aware that in some circles it is not orthodox to admit the possibility of alcohol being serviceable in any quantity, to any individual, in any condition of body or mind. To this statement I cannot assent. It is pernicious in itself, and doubly so in practice. Alcohol is a cardiac stimulant, of great value, and in cases of failing force, of heart debility, of sinking or of syncope, it is not only admissible, but it is demanded. Such is the daily experience of physicians the world over. If, therefore, the patient is prostrated; has a feeble pulse,

---

* When the stomach is irritable minute doses of concentrated food, for example, a teaspoonful of beef essence, repeated every fifteen minutes, will produce quiet and sleep.

a cold and clammy skin, a trembling tongue, and hurried breathing, the indication for the use of alcohol is clear and positive. The indications to cease its administration are, a return of warmth to the surface, the ability to retain and appropriate food, and a steady heart. Then total abstinence is as imperative, as was its opposite in other conditions. One fact I have noticed after years of close observation with all forms of alcoholism, which I desire to emphasize: I never saw delirium tremens supervene under the gradual diminution of alcoholic beverages, and never had occasion for padded room or physical restraint, in delirium tremens under the alcoholic treatment. Delirium tremens often occurs as the result of shock to the nerve centres, by the sudden and immediate abstraction of all stimuli, as is observed in prisons and hospitals whose practice favor that plan, but by the gradual removal of accustomed stimuli to avoid shock, never, in my observation.

In the discussion of treatment, so far, we have not gone outside the home, and in reviewing what has been written, it may be concisely stated thus. Alcoholic intoxication is a neurosis, which must be placed under the care of the family physician, and dealt with on the same principles as other disorders of the same class. The hospital or asylum treatment will be considered in another section.

DIFFERENT ALCOHOLS AND THEIR EFFECTS.

The opinion that alcohol, being a poison, should not therefore be used in the treatment of poisoning by its use, is commonly accepted as a truism, and bears upon its face the appearance of fairness, and reasonable argument. I shall devote a few moments to the consideration of alcohols, and I, think, demonstrate the fallacy of the popular opinion.

Modern chemistry has discovered no less than half a dozen varieties of alcohols, that are considered as important, each differing from the others in its atomic composition, and varying to some extent in results.

Modern pathology, too, has traced striking differences in the effects of these different varieties. So that we are not always certain whether the symptoms we observe, are the result simply of alcoholic poison, or of some of the new properties of alcohol, which disclose themselves in the various changes the drug undergoes after it enters the stomach.

In France a decided difference has been observed between the symptoms of the alcoholism of the country peasant, and that of the city workman, which difference is attributed to the difference in the properties of the alcohols consumed.

We are indebted also to Dr. T. L. Wright, of Bellefontaine, Ohio, for valuable observations on the different effects, not only of different alcohols, but of the same drug in its primary and its secondary results upon the same individual, which observations have great practical value. "When the operation of the stimulant is recent," says Dr. Wright, and it acts upon the "unchanged nervous system," the impression is quite different from what it is when the *chemical effects* of alcohol begin to declare themselves—these chemical changes being the elimination of other poisons, such as "carbonic acid and urea," which, entering the blood, reach the brain, especially at the end of a prolonged and profound debauch, when there is a complete change in the symptoms. In the beginning of a debauch the unchanged alcohol produces its specific effects, with which all are familiar, such as an exhilaration of the natural qualities of the subject of its use, exhibiting traits of character which are recognized as belonging to his individuality, etc., but continuing its use, as new qualities are set free, they invade the brain, until the whole nature seems to be changed, and as Dr. Wright forcibly expresses it, these natural traits and exhibitions of character are "lost and overborne by the tremendous and universal oppression of a new group of poisons, differing totally in their effects from alcohol." This "new group of poisons" is wrought out of the vital laboratory, in which are produced "chemical reactions of alcohol" which usurp the place of the original poison, and produce such changes of tissue and function, as are quite different from the effects of the drug before it has undergone chemical change, and thus eliminated additional forces, the effects of which vary.

M. Rabuteau, of Paris, demonstrates, in his "Elements de Pathologie," that alcohols are dangerous in proportion to the complexity of their atomic composition, and recent experiments upon animals have shown that certain varieties of alcohol, will not produce convulsions, for example, while others will, and that the effects on man should be studied with reference to this, and other newly discovered facts.

In a paper recently read by Professor Verga, before the Lombardy Institute of Science and Letters—

"He pointed out that the abuse of spirits is much more injurious than the abuse of wine, and related the case of a brandy manufacturer who, through the inhalation of the alcoholic fumes, in which he was obliged to spend much of his time, fell ill, and died of alcoholism, in a lunatic asylum. A distinction, too, should be drawn between old spirits and new spirits; between that obtained by distillation from wine and that obtained from grain."

In Germany a curious fact has recently been noticed : that symptoms of sleeplessness, excitement and delirium, resembling some forms of inebriety, are developed by the consumption of diseased potatoes and sausages. So close is the resemblance that the terms potato and sausage intoxication are applied to these symptoms, thus demonstrating that the blood contaminated by such poisons may derange the brain in like manner with alcohol, producing similar symptoms of intoxication.*

Dr. J. Milner Fothergill, of London, has shown that certain foods which are taken with apparent impunity

* British Medical Journal, Dec. 30, 1882.

into the stomach, really make no morbid impression during primary digestion, but during the stage of digestion when new products (poisons) are eliminated, which, entering the circulation, and lodging in the brain, disclose symptoms of mental alienation. This cannot be from the food in its original forms; but from the disintegration of its atomic qualities, and the chemical reactions consequent upon such process, new poisons are eliminated, and bring about symptoms of insanity or inebriation.

. So, also, the accomplished Dr. Raynor, of the Hanwell Asylum, in England, confirms these views. He laid before the International Medical Congress, in London, 1881, a paper on "Gouty Insanity," showing that the poison of gout produces similar symptoms, which vary in degree, in proportion to the intensity of the poison. He shows that when the poison is *not intense,* it results simply in sensory hallucinations or melancholy. When it is "*intense and sudden*" the result is mania or epilepsy. When " *intense and protracted,*" the usual result is general paralysis, and when attacking a person broken down by alcoholism or lead-poisoning, the result is " varying degrees of dementia."

Dr. Robert Grieve, of the Colonial Lunatic Asylum at Berbice, adds his testimony in the same direction, by affirming that "of the Coolie patients in that Institution, a very large proportion owe their mental troubles to the abuse of Cannabis Indica, or *bang,* and only a very few to alcoholic excess." *

* *British Medical Journal,* Dec. 30, 1882.

These are curious and interesting facts, and go to show how little we have known in the past, about the effect of alcohol upon the brain, and to open the door to much grander fields of research. We have learned this much, that the violent and abusive behavior of the drunkard, toward those especially whom he should love and cherish, is due to a change of cerebral function, which depends upon one or more of the "new group of poisons," described by Dr. Wright, who says, "There is a reason for this conduct quite different from the inborn wickedness to which it is usually ascribed."

These new poisons have called forth from the brain, passions that have been concealed or subdued, or originated by metamorphic change, and forced them into extravagant and violent forms of expression.

In the medical treatment of these cases, some of the symptoms of which are very similar, there should be no room for much divergence of opinion. If we have great feebleness and depression, feeble and rapid pulse, muscular tremor, cold and clammy skin, soft and tremulous tongue, with perhaps hallucinations of one or more of the senses, we want to bridge over a serious crisis, to save life long enough, to secure the acceptance of food, and other supporting treatment. We know that alcohol has been indicated in such cases for centuries; we know, in our own experience, that it has carried our patients over many a dangerous crisis, and we have no cause to stop and speculate as to the poison which has produced this condition. We know that alcohol will steady the heart, slow the pulse, warm the skin, and calm excitement, and we ought to use it. We should also know when to stop using it, and when to insist upon total abstinence from all intoxicants forever.

THE PSYCHOLOGY OF INEBRIETY.

Psychology has much to do with inebriety, not only as to its history, but as to its management. The career of an inebriate is distinguished by remarkable psychic exhibitions, which, if not understood, often lead to misapprehension and mismanagement. Though inebriety is a physical disease, it is nevertheless clearly identified with psychical phenomena, the study of which tends to a clearer apprehension of mental states. If we have a sensory or psychological effect, following any given state, or if we have a physical condition, that can be traced directly to sensory or psychical causes, we have a matter of great interest, which we cannot afford to overlook.

Reference has been made to "psychical traumatism," as an exciting cause of paroxsyms of inebriety. Such a term includes what many persons call mental shock, as sudden alarm, profound grief or sorrow, and surprise in a variety of forms; but there is a kind of physical impression, that is not shock nor surprise, neither grief nor joy, that may be often observed.

The bodily force of an habitual toper is, of course, weakened by his mode of life, and with it the mental functions share in a degree, either of feebleness, or of a departure from their normal parallelism with healthy organic life. It may not be the feebleness of dementia, such as so often attends paralysis, but it shows itself in waywardness, in a want of fixity, and a lack of power to coördinate itself with natural functions in other parts.

It may be that there is a sluggishness in all the activities of the being. Not so much a dullness of comprehension, perhaps, as a slowness to respond to suggestions, or appeals, which would be spontaneous and instinctive, if there was no disturbance of correlative forces. There can be no enthusiasm in such a condition. But let such a sluggishness be temporarily aroused, re-awakened, even by an artificial stimulant, and whether it be an arousement of psychical energy, which is transmitted to the consciousness or representation, or by a combination of other forces or qualities, the fact is still apparent, and is vivid with instruction.

Nothing is more common in the career of the inebriate class, in connection with psychical manifestations, than for them to recall, and relate, facts and incidents, while under the exhilarating influences of alcoholic beverages, that have lain dormant, and been forgotten beneath the heap of benumbed qualities, that are the product of alcoholic indulgence.

It is thus that men of this class so frequently, when in company with each other, as the intoxicating cup passes from lip to lip, and the tide begins to rise, find themselves revealing the background of life's picture, and bringing out qualities, and figures, that they usually desire to conceal. And thus, too, that after the period of exhilaration shall have passed, and they fall back upon the barren heap of their own depleted and exhausted energies, there shall be no vivid recollection of the descriptions they have given. What that background may be, what facts and figures it may reveal, of course, depends upon the temperament, taste and training of each individual.

For example. A group of such men, who have had a sound religious training, of similar social positions and associations, but who have been led into excess, and its accompanying evils, sit down together, for a game of some kind. The bottle is passed from one to another, and the whole circle presently feels the glow of excitement, which is arousing the dormant faculties of each; the memory is stimulated, and one of the party announces that, in his opinion, this game is wrong, and should not be continued. With this announcement, he repeats the instructions of his early life, and brings forward the counsels of those who have guided him. The others do the same, the game is laid aside, for the time, and a discussion is commenced. It would surprise many a stalwart temperance advocate, to listen to the theological disquisitions which flow from the lips of such a group, and how, one by one, they bring out the figures of the past, and array them appropriately on the face of the mental canvas, that they are exhibiting. Like an old painting with wrinkled canvas, faded outlines, and figures indistinct, when subjected to the fresh touch of the artist, his brush removes the cloudy surface, restores the figures, and brings out the whole in plain relief, that it may be examined and appreciated. It may not seem appropriate to give to the bottle, the attributes of the artist, but the illustration is by no means overdrawn. Whatever may be the deepest, and perhaps the remotest impression, or the latent quality of the mind, that belongs to its nature, or the accumulated gifts of culture or taste, they may all be obscured and hidden, behind the tarnish of time or abuse; but when the adventitious gloom is removed, the original, though it may be stained, and even torn, stands out, true to itself, and to its grouping with others. The subjects are various; Science, Politics, or any other, that may have had the most prominent place in the mind in times of sober living. The period is reached, however, in due course, when the state of mere elation or exhilaration, gives way to stimulation, by the continued use of the bottle; when the outlines of the reproduced figures become blurred again, and confused conversation takes the place of that which was clear and deliberate; the quiet discussion ceases, and a boisterous, perhaps a tragic end, may terminate the interview. The pleasing glow of excitement, resulting from the moderate use of the bottle, is lost by the stimulating effect of its contents, and the fresh and enjoyable reminiscences of the past, that cheered and enlivened their social intercourse, is lost in the extravagant and senseless jargon of the debauchee.

The lessons to be learned by these experiences, are instructive to those who have to do with inebriates, or to those who desire earnestly and honestly to study them.

I cannot well forbear the introduction at this point of a case, to illustrate the psychical condition, which attaches to some cases that are not traumatic, and yet at the same time are impressive, as facts.

Soon after taking my seat in a street car, in Philadelphia, about two years ago, a young man entered, but in a few moments changed his place, to a vacant seat at my side. As he did so, he mentioned my name, and presented before me, an envelope with his own address upon it, which I at once recognized, and greeted him accordingly. He was a former inebriate patient, whom I had not seen for eight years, or more. He was able to say that he had not tasted alcoholic liquors during that period; but he wished to narrate certain experiences that he had encountered during those years. He said, "I do not find it so hard to abstain from drink, as I do to avoid the conditions and circumstances which lead to drinking." Meeting with certain people, passing or frequenting certain places, and doing things without drinking, that he always used to do while drinking, was the hardest struggle of all. To be invited to drink by a stranger, or in a new place, was no inducement to indulge.

He narrated the following, to confirm his statement:—

He and a few comrades were in the habit, at a given hour in the morning, during the summer months, of visiting a certain saloon, where they partook of a fancy drink of rare attractions, the bartender being regarded as an expert in its manufacture. He said, "I seldom think of the place now, unless I happen upon it in my drives about the city, or meet with one of my old companions, during the hot months; but if the time and conditions are suitable, and I come within sight of it, instantly the figure of the expert barkeeper, the crackling ice, the cooling beverage, and the familiar forms and voices of my old friends, come vividly before me as a living picture, and I seem to be one of the old party. The impulse to enter is almost overpowering. The only thing I can do, (and I am thankful that I do it instinctively) is to put whip to my horse and drive away, at a rapid gait. Lest I may meet some of my friends on their accustomed route, I hasten toward the streets that are out of their familiar course.

The picture follows and haunts me, however, till I am in the Park, or out of the neighborhood, or at my stable. In this case, there was no traumatic shock, in the true sense, no sudden grief, or sorrow, or alarm, to throw him off his balance, nothing even to entice the old passion, but the simple association of thoughts and memories, clustered about a place, and a few persons, at a particular hour in the day, at a particular season of the year, producing a psychical impression, as difficult to overcome, as the group of traumatic effects described by Crothers and others.

Again. During his years of indulgence, there was a certain day, memorable as an anniversary of some special event, on which he opened for the first time a bottle of wine at dinner, of rare quality. After abandoning his drinking habits, at one of these anniversary days, which were still celebrated, he was suddenly impressed, while at the dinner table, with the thought, that the occasion would not be complete without the presence of that particular brand of wine, and proposed to his wife to send out and get a bottle.

She recognized, at once, the coincidence between the special day, and the special brand, and such an unexpected, and for the moment, alarming proposition, and as they looked at each other earnestly, he recalled, the proposition. But the recollection of the wine having been used, in commemoration of an event that he did not wish to cease celebrating, lingered in the mind, and tarried long enough to annoy him, and awaken a train of phenomena, which were symptomatic of nerve disturbance, to a degree to develop the prodromic signs of an invasion of the old disease. For several days, at the five o'clock dinner, these nervous symptoms were repeated, and nothing but a change of the hour to six, and going to bed soon after, for several consecutive evenings, allayed the irritability, and enabled him to recover his wonted steadiness of nerve; there was no longing for liquor, as in the chronic toper, no gastric yearning or sense of physical need. It was simply the association of thoughts and memories, in connection with an event which had not been completed by the omission of a conspicuous feature of the social entertainment, which on previous occasions, had been regarded as important, if not essential. No doubt that if the bottle had been presented, as a completion of the outfit, his will would have been captured, and he, overpowered.

So had the eight years been spent, since he had abandoned a life of excess. A young, active, prosperous business man, in daily conflict with a constitution, that

12

is signalized by a remarkable appetency to psychical manifestations.

Such protracted conflicts, when victorious, are deserving of honorable record, and when followed by defeat, should not be remembered with reproach, or condemned as types of unworthy manhood.

But few persons pause to think of the inner movings and impulses, the secret forebodings, of those who come into the world weighted with psychotic conditions, that predispose them continually to disaster and failure, and but few, who find themselves on an even plane of life, without such marked predispositions, are able to comprehend, much less to judge, their fellows.

In the case of this young man, as in other cases of excess, there were certain molecular changes which had been going on during the stage of his life that was given to dissipation, of which he was unconscious, that were imperceptibly laying the foundation for future functional disorder, and perhaps organic change, that magnified and intensified his proneness to morbid feeling and pursuit. While apparently healthy, and able to conduct a responsible business, it is at the expense of a constant strain, which in its turn, draws largely on the stock of psychical energy, that, together with the expenditure of physical strength, is calculated to keep him in a state of uncertainty,—all the time on the alert, living a life of intensity, and but for the remarkably developed capacity, either to adapt himself to new conditions, or promptly to evade or escape them, not by resistance, but by seeking adverse avenues or outlets, he would have been a wreck before now.

Had he fallen in with his former companions, and joined them at the bar of the saloon, it would have been impossible for him to have escaped indulgence, and its consequences—a debauch; his will would not have held the supremacy; but by adroitly averting the impending risk, he was safe. So with the bottle at dinner. Had he placed himself face to face with it, and attempted to antagonize its use, by his will-effort merely, he would have yielded, been debauched, and awakened from it, to combat life again, possibly, with an added lesion, either of function or structure, fastened upon some portion of his complex organism, thus increasing his inability to resist or evade future attacks.

He did not drink, because he was able to disconnect the chain of circumstances and associations formerly connected with the act. Had he caused them to unite, and thus made the links consecutive in his mind, it would have seemed to him necessary to drink. There are many men who scarcely think of drinking, aside from the associated facts and conditions, which are, as it were, cartooned in striking outline before their mental vision, and if the picture is not complete, if the central figure in all circles and associations of drinking men—the bottle—is not present to the mental vision, there is not much risk. The bottle is the pivot on which everything turns, and around which everything revolves. Repudiated, as it should be, as essential to any civilization, it continues to hold its conspicuous position, and challenges the interest of its devotees to a degree, that efforts to restore them to society and themselves are utterly fruitless, while it continues to hold its supreme attitude, in relation to their lives.

As association has much to do with continuing a habit, so change of association, has much to do with a change of habit; and as habit so often depends upon abnormal states of the physical system, so avoidance of habit, will frequently relieve the organs of the body from service that is constantly demanded, by its morbid conditions, which are symptomatic of disease. Hence the advantage of change of residence for a time, to get beyond the voice and influence of familiar associations. Among new places, and new faces, new impulses begin to throb, new thoughts to form, new remedies to be employed, and a new life to begin, with new experiences to strengthen its beginning, until a new constitution and a new character, demonstrate that transplanting to a new soil, is the first movement towards harvesting new, fruit.

Such transplanting is, however, impracticable, in the majority of cases, unless, indeed, the Government should discover that it is the duty of the State, to protect and restore such cases, as she does her insane, her blind, her mutes, and as she does, in another manner, her wayward and criminal classes.

When the people shall learn that insanity, idiocy, epilepsy, inebriety, and, to a great degree, crime, all belong to the same family, and all grow in the same soil, then will provision be made alike for each.

But there is no class of citizens in the Commonwealth more isolated from the guidance and protection of law, from the fostering influence of domestic and social attachments, from the relief that comes from a share in the common bond that unites men, in the relation of fraternity, and from the church and her appliances, than the inebriate class. This is not surprising, however, because there is no form of humanity more loathsome, than that which is disfigured, and spoiled by the toxic power of alcohol. Hence, as a class, they are drawn towards each other by affinities, that are common and peculiar to themselves. Having no union of tastes or tendencies, with the multitude of men, who crowd the same roadway of life, they diverge into paths distinctively their own. In consequence, too, this very separation from others, and this natural grouping together of themselves, in their own morbid ways, causes them to move along with more friction, than other classes of men. And hence they become the victims of each other's cupidity and deceit, and their pilgrimage is hard and full of misery to themselves, and a source of suffering, distrust, fear and expense to society.

The illustrations already offered, exhibit the operation of psychic force upon bodily functions, or the reverse. In the class of persons we are dealing with, there are numerous and constantly occurring phenomena of this character, which have much to do in guiding the treatment, but they will not be available for practical application, until this whole subject is withdrawn from the domain of morals, as the starting point of investigation and discussion. The soil in which it is seeded, and where it takes root, is in the realm we have been traversing, in which the moral aspect is simply a secondary correlative. At this point, the natural enquiry is for the evidences of physical disease, seeing that throughout this narrative, there seems to be no period, when the craving or passion for drink was prominent. In answer, it may be said, referring to the case noted, that after eight years of sober and industrious living, with the discipline and culture of such living, the attitude of the man's thoughts and feelings, in their relation to his previous dissipations, was entirely changed, and his approaches toward them were necessarily in the order of his reversed thoughts. The tendency, however, the predisposition, the susceptibility which constituted the pre-existing abnormal condition, though latent and remote, was still there, and it was the struggle to reinstate the old order of things, and to bring out the latent morbid element, and give it the dominant position in his case, that caused the train of nervous symptoms to be set up, which threatened to overcome and prostrate him.

Morbid psychology embraces an almost limitless roll of, as yet, nameless phenomena, which will, in time, naturally group themselves into classes, and take their places as symptoms of abnormal conditions, which are now only recognized as eccentricities, and peculiarities, that appear rather as accidents of character, than as the legitimate results of pathological states.

Psychological science is constantly unfolding secrets, that have been hidden by the ignorance of the past, and especially in the field we are now exploring. Witness the illustrious services of modern psychologists, both at home and abroad, whose works are doing so much to instruct and modify public sentiment, by widening and deepening its scope of observation, and extending its capacity to know, and to judge of human conditions, and conduct.

The psychcial side of the inebriate, is a most profound and absorbing study. It is a feature in his career which must be˙ understood, if we would rightly understand him.

Elsewhere, it has been seen how mental shock has been the exciting cause of drunkenness. Indeed, nothing is more common, than for men to excuse themselves for drinking to excess, because of some disappointment, misfortune, or grief. Financial embarrassment, loss of friends or kindred, are conspicuous among the exciting causes of excess; and in this regard there is a singular conformity with the causes of insanity, as laid down in the reports of asylums for the insane.

It is equally true, however, that shock, fright, or other sudden emotion, may act as an immediate cause, or starting point of recovery. Dr. Rush mentions the story of several young men who became intoxicated in a little cabin, located on the banks of the James River, in Virginia, but a sudden rise in the river came on, and floated the cabin and its contents upon the dangerous current, and the peril of their situation, when they began to apprehend it, was so alarming to them, that they were entirely sobered, by the time they were driven ashore. He mentions also, on the authority of Dr. Witherspoon, the case of a Scotchman, who was always cured of a fit of drunkenness by being made angry, and he was always angered, by talking to him against religion.

From my own clinical records I could narrate case after case, in which mental impressions had much to do with modifying the action of intoxicants, and in some instances, where prompt recovery, not only from a single attack, but from future excess, was induced by strong and sudden mental emotion. A young man of culture and wealth, whose opportunities in life were exceptional, had for years given himself to loose company and dissipation. Every motive had been appealed to by his parents, and every inducement offered, for him to make an effort to change his course, but without avail. One evening, on returning home, in a state of only moderate intoxication, his father directed him to go to his room, pack his trunk, and be ready for a start by a morning train, for an institution for inebriates. Looking at his hitherto forgiving and indulgent father, he put the question, "Are you in earnest?" The reply was prompt and decisive, "Yes, in earnest."

The following morning found the young man at the breakfast table, with his mind impressed as it never had been before. He had met a crisis that he had not for a moment anticipated. He saw himself in a new light, even as a public inebriate, with a father whom he had always trifled with, but who was now stern, decided and inflexible. The attitude of his home, and its inmates was changed toward him; his own attitude in relation to them and to society was modified. A new set of impressions had established themselves in his mind, and, for the time, were supreme in their influence upon his consciousness, his judgment and his will.

They overpowered him, and he determined to make an effort at self-reclamation, which was for a considerable time successful.

This sudden and unlooked-for shock, affected, not only his moral nature, arousing him to a consciousness of his relations to others, such as he had not before thought of, but upon it, supervened a change of physical state.

The deepest emotions being stirred, a new power was felt in the whole being, and all the functions were re-animated, so that the man felt himself new again, and while the overpowering influence of the newly stimulated emotional nature, maintained the supremacy, so long did he feel secure from relapse. In the case of this young man there was first, a new awakening to his true condition. His pride was then stirred, to avoid the disrepute of a committed inebriate. His fear of punishment, of being banished from home, next asserted itself, and thus, for a time, his morbid passion for drink, was held in abeyance.

It is in obedience to the same law, that so many sudden "reformations" as they are called, appear, in times of great public interest and effort in this behalf, and for similar reasons, that such "reformations," are transient and unstable. The same thing is seen in other maladies.

But a few months since I was called to see a young

lady of refined feelings and nature, who was suffering from an attack of acute mania. It was the third time I had treated her, during the last few years, for the same affection. She was kept at home. Her father had spent the last year or more of his life, in an asylum for the insane. She visited him during his residence there, was familiar with the place, and had frequently spoken in praise of the institution, and its management. During this attack, which was a little more obstinate and protracted than the others, I recommended to her family her removal to the same institution. It was mentioned to her, and she consented. Her family were desirous that it should be done. The necessary papers were executed, and a day fixed for her departure.

During the preparation for her removal, her whole thought seemed to be absorbed with the question how her mother, who was in feeble health, could get on without her presence and assistance. This was the preponderating thought that occupied her mind, day and night, and in trying to solve it, she was so much improved when the time came for her removal, that she accompanied her sister to my office, for the purpose of satisfying me, that there was no need of her going, and she remained at home and is now as well as usual. The one absorbing thought of her feeble mother was the preponderating and controlling thought. It possessed her, and took the place of the morbid fancies and delusions, that were urging her to self-destruction.

Such cases as these are constantly observed, and nowhere more conspicuously than among inebriates. The shock of an accident, or of a new and sudden change of base, with reference to any given purpose or plan, will give a new direction to the life, by its effect on the animal functions.

No fact in physiology is better known, perhaps, than that organic functions may be materially modified and even suspended by the emotions. All the processes of nutrition, acted upon by the vascular and nervous systems through the emotional nature, are subject to striking modifications: and this fact should be availed of, in conceding the peculiarities of inebriation, as they are manifested in different individuals. With such knowledge, and the practical application of such principles, to the treatment of inebriety, there will be much more hope of success than by the present popular mode of promises and pledges, seventy-five per cent. of which are, in the very nature of things, broken.

SUMMARY.

In concluding these pages, it may assist the reader to recapitulate, in a brief summary, the chief points to be remembered, in connection with the doctrine of disease as applied to alcoholic intoxication.

1. Some persons are born with an alcoholic diathesis, that is to say—with an appetency for alcoholic beverages, and a tendency to intoxication.

2. The desire for such stimulants is from within, its source and origin being independent of external temptations.

3. It may be prematurely evolved by a careless indulgence in intoxicating liquors, without its possessor being aware of its existence.

4. It is the internal craving for alcoholic liquors, and for their intoxicating effect, that constitutes the disease, and not the fact of drunkenness.

5. It may be produced by external injuries, especially. of the head, or by other sudden, alarming or depressing shock to the nervous system.

6. It may manifest itself, in the form of unexpected paroxysms, with intervals of sobriety, and even of dislike for liquors.

7. These intervals are of varied duration as to time, but when the paroxysms occur, no considerations of home, or duty, or affection, or morals, or religion can influence the victim to pursue a different course.

8. The paroxysms are apt to increase in frequency and duration with their repetition, so that chronic

alcoholism may supervene, and end the career of the inebriate.

9. Frequently there are well marked premonitory signs, which introduce a paroxysm of intoxication. Such as restlessness, irritability, general *malaise*, or it may be, the occurrence of an injury.

10. The paroxysms may be arrested, prevented, or controlled, by becoming familiar with the prodromic symptoms, and giving timely heed to their admonitions by the use of remedial measures.

11. They may terminate at a certain period of life, by the apparent exhaustion of the hereditary taint, at what is called the climacteric period.

12. The hereditary tendency may be transmitted in a direct line of inebriety, or it may be deflected, and appear by transmutation, as epilepsy, insanity, chorea, hysteria, or even—crime.

13. While there are resemblances between inebriety and insanity, in some of their manifestations, there are decided differences in their pathology, which entitle them to be regarded as distinct, though somewhat similar diseases.

14. In some inebriates there is a state of partial unconsciousness during a debauch, in which they may commit acts of violence, and have no memory of it afterwards; a state that is called cerebral automatism, or trance.

15. The most successful treatment is that which combines with wholesome restraint, psychological and hygienic methods, such as are successfully employed in hospitals for inebriates.

INDEX.

183

13

BLAKISTON, SON & CO., 1012 Walnut Street, Philadelphia, call attention to their large stock of books on Medicine, Dentistry, Pharmacology, Popular Science, Hygiene, Chemistry, Microscopy, Nursing and Invalid Cookery, Hospital Management, etc. In making a special study of this branch of book-making and selling, they can always give the latest information in regard to new books and other publications.

☞ *The following is a select list of Books on Popular Science, Health and Hygiene. Full Catalogues sent upon application.*

THE AMERICAN HEALTH PRIMERS. Edited by W. W.

Keen, M.D. Bound in Cloth. Price 50 cents each. Paper covers 30 cents each.

PRICE, SECURELY BOUND IN PAPER, 30 CENTS.

I. Hearing and How to Keep It. With illustrations. By Chas. H. Burnett, M.D., of Philadelphia, Aurist to the Presbyterian Hospital, etc.

II. Long Life, and How to Reach It. By J. G. Richardson, M.D., of Philadelphia, Professor of Hygiene in the University of Pennsylvania.

III. The Summer and Its Diseases. By James C. Wilson, M.D., of Philadelphia, Lecturer on Physical Diagnosis in Jefferson Medical College.

IV. Eyesight, and How to Care for It. With Illustrations. By George C. Harlan, M.D., of Philadelphia, Surgeon to the Wills (Eye) Hospital.

V. The Throat and the Voice. With illustrations. By J. Solis Cohen, M.D., of Philadelphia, Lecturer on Diseases of the Throat in Jefferson Medical College, etc.

VI. The Winter and Its Dangers. By Hamilton Osgood, M.D., of Boston, Editorial Staff Boston *Medical and Surgical Journal.*

VII. The Mouth and the Teeth. With illustrations. By J. W. White, M.D., D.D.S., of Philadelphia, Editor of the *Dental Cosmos.*

VIII. Brain Work and Overwork. By H. C. Wood, Jr., M.D., of Philadelphia, Clinical Professor of Nervous Diseases in the University of Pennsylvania, etc.

IX. Our Homes. With illustrations. By Henry Hartshorne, M.D., of Philadelphia, formerly Professor of Hygiene in the University of Pennsylvania.

X. The Skin in Health and Disease. By L. D. Bulkley, M.D., of New York, Physician to the Skin Department of the Demilt Dispensary and of the New York Hospital.

XI. Sea Air and Sea Bathing. By John H. Packard, M.D., of Philadelphia, Surgeon to the Episcopal Hospital.

XII. School and Industrial Hygiene. By D. F. Lincoln, M.D., of Boston, Mass., Chairman Department of Health, American Social Science Association.

This series of American Health Primers is prepared to diffuse as widely and cheaply as possible, among all classes, a knowledge of the elementary facts of Preventive Medicine, and the bearings and applications of the latest and best researches in every branch of Medical and Hygienic Science. They are intended incidentally to assist in curing disease, and to teach people how to take care of themselves, their children, pupils, employés, etc.

They are written from an American standpoint, with especial reference to our Climate, Sanitary Legislation and Modes of Life.

Select List of Books.

"As each little volume of this series has reached our hands we have found each in turn practical and well-written."—*New York School Journal.*

"Each volume of the 'American Health Primers' *The Inter-Ocean* has had the pleasure to commend. In their practical teachings, learning, and sound sense, these volumes are worthy of all the compliments they have received. They teach what every man and woman should know, and yet what nine-tenths of the intelligent class are ignorant of, or at best, have but a smattering knowledge of."—*Chicago Inter-Ocean.*

"The series of American Health Primers deserves hearty commendation. These handbooks of practical suggestion are prepared by men whose professional competence is beyond question, and, for the most part, by those who have made the subject treated the specific study of their lives. Such was the little manual on 'Hearing,' compiled by a well-known aurist, and we now have a companion treatise, in *Eyesight and How to Care for It,* by Dr. George C. Harlan, surgeon to the Wills Eye Hospital. The author has contrived to make his theme intelligible and even interesting to the young by a judicious avoidance of technical language, and the occasional introduction of historical allusion. His simple and felicitous method of handling a difficult subject is conspicuous in the discussion of the diverse optical defects, both congenital and acquired, and of those injuries and diseases by which the eyesight may be impaired or lost. We are of the opinion that this little work will prove of special utility to parents and all persons intrusted with the care of the eyes."—*New York Sun.*

"The series of American Health Primers (now entirely completed) is presenting a large body of sound advice on various subjects, in a form which is at once attractive and serviceable. The several writers seem to hit the happy mean between the too technical and the too popular. They advise in a general way, without talking in such a manner as to make their readers begin to feel their own pulses, or to tinker their bodies without medical advice."—*Sunday-school Times.*

"*Brain Work and Overwork,* by Dr. H. C. Wood, Clinical Professor of Nervous Diseases in the University of Pennsylvania, to city people, will prove the most valuable work of the series It gives, in a condensed and practical form, just that information which is of such vital importance to sedentary men. It treats the whole subject of brain work and overwork, of rest, and recreation, and exercise in a plain and practical way, and yet with the authority of thorough and scientific knowledge. No man who values his health and his working power should fail to supply himself with this valuable little book."—*State Gazette, Trenton, N. J.*

"An unexceptional household library."—*Boston Journal of Chemistry.*

"Every family should have the entire series; and every man, woman, and child should carefully read each book."—*Alabama Baptist.*

"Everybody knows that it is uncomfortable to be cold, but few know that undue exposure to cold shortens life, and still fewer the nature of the safeguards that ought to be taken against it. . . . This little book, *Winter and Its Dangers,* contains a mass of well digested and practical information."—*St. Louis Globe Democrat,* in a two-column review, Nov. 24th, 1881.

"The whole series is a particularly useful one. and should be added to the reference books of Academies and High Schools."—*Zion's Herald, Boston.*

LONG LIFE. The Art of Prolonging Life. By C. W. Hufeland. New Edition. Edited by Erasmus Wilson, M.D. 12mo. Price $1.00

"We wish all doctors and all their intelligent clients would read it, for surely its perusal would be attended with pleasure and benefit."—*American Practitioner.*

"We all desire long life, and the attainment of that object, as far as it can be accomplished by an adherence to the laws prescribed by nature, may be furthered by a perusal of Dr. Hufeland's book, which is written in a style so perspicuous and free from technicalities as to be readily comprehended by non-professional readers."—*Philadelphia Evening Bulletin.*

"The work is a rational and well ordered presentment on a subject of moment to all. It prescribes no panacea, but puts in requisition instrumentalities that are in every one's reach. It should be read by all."—*North American.*

ALCOHOL AND TOBACCO. Alcohol; its Place and Power. By James Miller, F.R.C.S.; and, Tobacco; its Use and Abuse. By John Lizars, M.A. The two essays in one volume. Cloth, Price $1.00
Either essay separately. Price 50 cents.

"A perusal of this work rather startles a smoker and chewer, and gives one an idea of the silent work going on in the system. It certainly shows that a man must sooner or later feel the pernicious influences of alcohol and tobacco. Let smokers and absorbers read it, and then make their calculations on the length of time they will last under a continuation of the evils, and whether it is not best to heed the facts there laid down and 'moderate' a little."—*Californian.*

"They are full of good, strong, medical sense, and ought to be very influential agents against the vices they assault."—*Congregationalist.*

"We have seldom read an abler appeal against the demon of intemperance, or one enforced by more cogent arguments."—*Philadelphia Inquirer.*

THE MENTAL CULTURE AND TRAINING OF CHILdren. By Pye Henry Chavasse 12mo. Price $1.00; paper cover, 50 cts.

The mental culture and training of children is of immense importance. Many children are so wretchedly trained, or rather, not trained at all, and so mismanaged, that a few thoughts on this subject cannot be thrown away, even upon the most careful.

ON INDIGESTION. Indigestion: What It Is; What It Leads To: and a New Method of Treating It. By John Beadnell Gill, M.D. Second Edition. 12mo. Price $1.25

"Indigestion, pure and simple, is responsible for almost all the other diseases that flesh is heir to. Rheumatism and gout are the direct consequences of this disorder, as well as heart and lung troubles. To cure this diseased state of digestive and assimilating organs Dr. Gill, a distinguished English physician, has written this able treatise. He has summed up some eighty-eight cases and their natural remedies, besides a system of eliminants and tonics. Great stress is laid on proper bathing, as a curative agent, and on drinking hot water and its other uses. The fact of a second edition being required within a few months of the first, needs no comment, and points the demand."—*Philadelphia Ledger.*

HOW TO LIVE.

A GUIDE TO HEALTH AND HEALTHY HOMES.

By GEORGE WILSON, M.D.   Second Edition. Edited by Joseph G. Richardson,
Professor of Hygiene University of Pennsylvania.

314 Pages.   Price, cloth, $1.00; Paper covers, 75 Cents.

SCOPE OF THE WORK.

The object of the author in writing this book is to advance the art of
preserving health; that is, of obtaining the most perfect action of body
and mind during as long a period as is consistent with the laws of nature.
Though many books have been written analogous to the subject, there is
none like this; sufficiently simple, and at the same time systematic and
comprehensive. A glance at the table of contents will convince the reader
of its completeness and reliability as a guide to all those wishing to lead a
happy, healthy and long life.

Chapter I is a general introduction to the whole subject, giving a few
statistics in regard to death rates, and remarks showing the great number
of preventable diseases and the possibility of reducing the many early
deaths by a proper regard of simple health rules   Chapter II is explanatory of the different functions of the human body, for the more thorough
understanding of the following chapters. Chapter III is headed Causes
of Disease, self induced and social; treating of intemperance in food as
well as drink and tobacco, mental overwork, immorality, idleness, irregular
modes of living, sleep and clothing, contagious diseases, consumption. etc.;
unsound food, impure air, etc., etc. Chapter IV is more particularly
devoted to food and diet, their proper choice, digestive qualities and preparation.   Chapter V treats the subjects of cleanliness and clothing.   It is
astounding, the ignorance displayed by the majority of people on these
points, and Dr. Wilson gives many useful hints invaluable to every one.
Chapter VI is on Exercise, Recreation, etc., giving the proper amount of
exercise to be taken by boys and girls, young and old, explaining its
necessity and good effects; details are also given for the proper training
for racing and athletic sports as recommended at various universities.
Chapter VII treats of the more general theme of the Home and its surroundings, drainage, water supply, ventilation, warming, outside premises,
and innumerable hints of value about choosing or building a new home, and
the alteration and healthful arrangement of an old one.   Chapter VIII,
Diseases and their prevention, and concluding remarks.

Only an outline of the scope of this book can be had from these few general headings, but it would be impossible to give in so limited a space the
thousand and one subjects handled, popular errors corrected, and useful
hints given by Dr. Wilson. in these three hundred and fourteen closely
printed pages. A general index completes the volume. and the well known
name of Prof. Richardson on the title page, as editor. is an additional guarantee of its trustworthiness as a guide in all things relative to health and
How we should live.

PRESS NOTICES.

" The book aims at the *prevention* of Disease.   It abounds in sensible
suggestions. and will prove a reliable guide."—*Churchman.*

" A most useful and, in every way, acceptable book."—*New York Herald.*

" Full of good sense and sound advice."—*Educational Weekly.*

" Deserves wide and general circulation."—*Chicago Tribune.*

SLIGHT AILMENTS:

THEIR NATURE AND TREATMENT.

By LIONEL S. BEALE, M.D.

Second Edition, Revised and Enlarged.

OCTAVO. PRICE, CLOTH BINDING, $1.25; BOUND IN PAPER, 75 CENTS.

Every one suffers from time to time with slight derangements of the health; derangements not dependent upon or likely to determine any important change in any organ or tissue of the body, but due to some temporary disturbance which, though painful and unpleasant, may be easily relieved by any one understanding their nature and cause.

A little too much food, or food of a bad kind, or food badly cooked, or eaten at the wrong time or too quickly, a glass of bad wine, bad milk or water, to say nothing of the disturbances occasioned by changes of atmosphere, and a hundred other causes, bring about such normal changes in the body as to make even the strongest and healthiest among us to feel for a time unwell; almost every one, in fact, experiences such departures from the healthy condition. It is about these Slight Ailments, which cause so much discomfort, and often a great deal of pain, that Dr. Beale treats. The old adage, that "prevention is better than cure," applies pertinently to slight ailments, as it is these which are often the forerunners of disease and doctor's bills.

OPINIONS OF THE PRESS.

"It abounds in information and advice, and is written for popular use."—*Philadelphia Bulletin.*

"A valuable work for the family library."—*Boston Transcript.*

"Clear, practical, and a valuable instructor."—*Baltimore Gazette.*

"In a very important sense, a popular book."—*Chicago Advance.*

"An admirable treatise upon the minor ills which flesh is heir to."—*Springfield Republican.*

Any book in this Catalogue will be sent, postage prepaid, upon receipt of price. See pp. 14 and 15 for SPECIAL OFFER. Money should be sent by Draft, Post Office Money Order, or Registered Letter.

The publishers have an extensive stock of books in all branches of Medicine and Science. Catalogues furnished upon application, correspondence solicited.

P. BLAKISTON, SON & CO.,

BOOKSELLERS, PUBLISHERS AND IMPORTERS,

1012 WALNUT ST., PHILADELPHIA.